Machine Learning, Blockchain, and Cyber Security in Smart Environments

Machine Learning, Blockchain, and Cyber Security in Smart Environments: Applications and Challenges provides far-reaching insights into the recent techniques forming the backbone of smart environments, and addresses the vulnerabilities that give rise to challenges in real-world implementation. The book focuses on the benefits related to emerging applications, including artificial intelligence, machine learning, fog computing, blockchain, and cyber security.

Key Features:

- Introduces the latest trends in the fields of machine learning, blockchain and cyber security
- Discusses the fundamental architectures and challenges of these concepts
- Explores recent advances in machine learning, blockchain, and cyber security
- Examines recent trends in emerging technologies

This book is primarily aimed at graduates, researchers, and professionals working in the areas of machine learning, blockchain, and cyber security.

Chapman & Hall/CRC Cyber-Physical Systems

Series Editors:

Jyotir Moy Chatterjee
Lord Buddha Education Foundation, Kathmandu, Nepal

Vishal Jain
Sharda University, Greater Noida, India

Cyber-Physical Systems: A Comprehensive Guide
By: Nonita Sharma, L K Awasthi, Monika Mangla, K P Sharma and Rohit Kumar

Introduction to the Cyber Ranges
By: Bishwajeet Pandey and Shabeer Ahmad

Security Analytics: A Data Centric Approach to Information Security
By: Mehak Khurana and Shilpa Mahajan

Security and Resilience of Cyber Physical Systems
By: Krishan Kumar, Sunny Behal, Abhinav Bhandari and Sajal Bhatia

For more information on this series please visit: https://www.routledge.com/Chapman--HallCRC-Cyber-Physical-Systems/book-series/CHCPS?pd=published,forthcoming&pg=1&pp=12&so=pub&view=list?pd=published,forthcoming&pg=1&pp=12&so=pub&view=list

Machine Learning, Blockchain, and Cyber Security in Smart Environments

Applications and Challenges

Edited by
Sarvesh Tanwar
Sumit Badotra
Ajay Rana

CRC Press
Taylor & Francis Group
Boca Raton London New York

CRC Press is an imprint of the
Taylor & Francis Group, an **informa** business

A CHAPMAN & HALL BOOK

First edition published 2023
by CRC Press
6000 Broken Sound Parkway NW, Suite 300, Boca Raton, FL 33487-2742

and by CRC Press
4 Park Square, Milton Park, Abingdon, Oxon, OX14 4RN

CRC Press is an imprint of Taylor & Francis Group, LLC

Library of Congress Cataloging-in-Publication Data
Names: Tanwar, Sarvesh, editor.
Title: Machine learning, Blockchain, and Cyber Security in Smart
Environments : Application and Challenges / edited by Sarvesh Tanwar, Sumit Badotra, Ajay Rana.
Description: First edition. | Boca Raton : Chapman & Hall/CRC Press, [2022] | Series: Chapman & Hall/CRC cyber-physical systems | Includes bibliographical references and index. | Summary: "Machine Learning, Cyber Security and Blockchain in Smart Environment: Application and Challenges provides a deep insight into the recent techniques which form the backbone of smart environment such as Artificial Intelligence, Blockchain, Cyber Security, Machine Learning and Fog Computing, and addresses the vulnerabilities that cause hindrance for the real-world implementation. This book is designed for university students and researchers to impart knowledge about cutting-edge technologies that deliver vast amount of information seamlessly and provides considerable visibility about real time problems that need to be addressed. The book focuses on the benefits related to the emerging applications such as machine learning, blockchain and cyber security. This book is primarily aimed at graduates, researchers and professionals working in the areas of machine learning, blockchain and cyber security"-- Provided by publisher.
Identifiers: LCCN 2022004727 (print) | LCCN 2022004728 (ebook) | ISBN 9781032146393 (hbk) | ISBN 9781032146416 (pbk) | ISBN 9781003240310 (ebk)
Subjects: LCSH: Artificial intelligence--Industrial applications. | Computer networks--Security measures. | Machine learning. | Blockchains (Databases)
Classification: LCC TA347.A78 M333 2022 (print) | LCC TA347.A78 (ebook) | DDC 006.3--dc23/eng/20220525
LC record available at https://lccn.loc.gov/2022004727
LC ebook record available at https://lccn.loc.gov/2022004728

ISBN: 978-1-032-14639-3 (hbk)
ISBN: 978-1-032-14641-6 (pbk)
ISBN: 978-1-003-24031-0 (ebk)

DOI: 10.1201/9781003240310

Typeset in Palatino
by SPi Technologies India Pvt Ltd (Straive)

Contents

Preface

Technology presents a golden opportunity for mankind to brighten the prospects of future generations. *Machine Learning, Blockchain, and Cyber Security in Smart Environments: Applications and Challenges* provides profound insights into recent techniques designed to augment machine learning, the Internet of Things, big data and cyber security in various applications such as healthcare, commerce, e-communications, agriculture, and the smart world. Our various environments, including homes and workplaces, require regular monitoring to manage and eliminate anomalies effectively. With the dawn of artificial intelligence (AI), human potential can be mimicked by deploying a variety of different algorithms. Machine learning, together with the use of statistical techniques, is an interdisciplinary field that enables computer systems to learn and adapt according to their environment. Virtual assistants like Alexa, Siri, and Bixby can respond to your queries, schedule appointments, book restaurant, play music, and control smart devices.

Cyber security is not only a booming research area within computer science but also plays a major role in the life of almost every individual, organization, society, and country. We live in an era that is increasingly reliant on technology. Sensitive information such as bank details, government identification details, and social security numbers are stored on web portals. Data breaches can lead to economic, reputational, and regulatory damage. Individuals and the corporate sector have to implement cyber security protocols to securely store their information. The popularity and utility of blockchain technology has significantly changed the working of digital transactions. With an increasing number of consortia, cheaper running costs, and more pilots and tests in progress, blockchain is rapidly becoming more widely accepted. Blockchain investment is used to provide trust in decentralized environments with relatively inexpensive security encryption, transparent transactions, and traceability. This book mainly focuses on the benefits of emerging applications such as machine learning, blockchain, and cyber security. We can make use of these applications for security, for learning and other things by enabling systems to automatically learn and improve from experience without being explicitly programmed.

This book aims to teach university students and researchers about cutting-edge technologies that can deliver a vast amount of information seamlessly, as well as investigating real-time problems that need to be addressed. The chapters deal with different aspects of applications and challenges in machine learning, blockchain and cyber security, including:

- Multidisciplinary applications of machine learning
- Machine learning in healthcare, Industry 4.0, agriculture and image processing
- Machine learning and cyber security
- Machine learning in medical robots
- Machine learning for data security
- Blockchain technology application platforms

- Blockchain in healthcare, supply-chain management, and government policies
- Blockchain in IOT and public key infrastructure (PKI)
- Blockchain for data security
- Cybersecurity testbeds, tools, and methodologies

Editors

Dr. Sarvesh Tanwar is an associate professor at Amity Institute of Information Technology (AIIT), Amity University, Noida, India. She is head of AUN Blockchain and Data Security Research Lab. She obtained her M.Tech (CSE) degree from MMU, Mullana, and PhD from Mody University, Laxmangarh. She has more than 15 years' teaching and research experience. Her research interests include public key infrastructure (PKI), cryptography, blockchain, and cyber security. She has published more than 80 research papers in international journals and conferences. She is currently supervising six PhD students and has previously supervised five M.Tech research students. She has filed 18 patents and one copyright. She is a senior member of IEEE, a life member, Cryptology Research Society of India (CRSI), Indian Institute of Statistics, Kolkata, India, and a member of the International Association of Computer Science and Information Technology (IACSIT), Singapore. She is a reviewer for *Journal of Cases on Information Technology (JCIT)*, *IEEE Access, MDPI, Asian Research Journal of Mathematics* and *Inderscience*, a member of the editorial review board of *IJISP* and IGI Global, USA. She is a member of the editorial board of the *International Journal of Research in Science and Technology (IJRSTO)*, Ghaziabad, UP, *Advances in Science, Technology and Engineering Systems Journal (ASTES)*, US and IAENG.

Dr. Sumit Badotra is currently an assistant professor in the School of Computer Science and Engineering, Lovely Professional University, Punjab, India. He has five years of research experience in software defined networks (SDN). During his PhD, he also worked as a junior research fellow on a project funded by the Government of India. He has published over 15 papers in SCI/Scopus/UGC-approved journals and over 10 papers in reputed national/international conferences/book chapters. He has published two patents and filed eight patents in his fields of interest. He is an active reviewer for a number of journals including *Supercomputing, IEEE Access, Cluster Computing, IJCS, and PLOS One*. He is currently supervising two PhD and two M.Tech students.

Dr. Ajay Rana, a renowned educationist, researcher, teacher, innovator, strategist, and committed philanthropist, is the vice chancellor of Shobhit University, Meerut. He previously served Amity University, Uttar Pradesh, Noida for more than 20 years as dean, director, professor in Computer Science and Engineering and senior vice president (RBEF – A Trust of Amity), Amity Education Group. He obtained his M.Tech and PhD in computer science and engineering. His areas of interest include machine learning, the Internet of Things (IoT), augmented reality, software engineering, and soft computing. He has more than 72 patents to his name in the fields of IoT, networks, and sensors. He has published more than 300 research papers in reputed journals and national and international conferences, co-authored eight books, and co-edited 36 conference proceedings. Dr. Rana is the founding chair of AUN Research Labs, executive committee member of IEEE Uttar Pradesh Section, Senior Member of IEEE, and life member of the CSI and ISTE. He is also a member of the editorial board and review committee of several journals. Dr. Rana possesses deep organizational ethics and believes in holding the hand of every individual who wishes to succeed in life. Dr. Rana has organized over 500 conferences, workshops, faculty development programs, seminars and talks sponsored by IEEE, Springer, CSI, and others. He has won 243 awards in recognition of his work in education and research including Corona Warriors Award 2020, Life Time Achievement Award, Frontiers of Knowledge Award, Man of Academic Advisory Award, Icon of Healthcare Award, Med Achievers, Leadership Award BRICS – MSME, Most Influential 100 Directors of India, Educlusion Award Singapore, IT Next CIO, International *WHO'S WHO* USA, Italian Machine Tool Technology Award, and many more. He has been awarded an honorary professorship at ULAB Bangladesh. He is currently a member of the Consultative Group NEP 2020 CBSE, Ministry of Education, Government of India.

Contributors

Riya Aggarwal
Amity University
Noida, India

Alpana
Manav Rachna University
Faridabad, India

Sumit Badotra
Lovely Professional University
Punjab, India

Amit Bairwa
Manipal University
Jaipur, India

Avali Banerjee
Guru Nanak Institute of Technology
    (GNIT)
Affiliated under Maulana Abul Kalam
    Azad University of Technology
    (MAKAUT)
West Bengal, India

Salil Bharany
Guru Nanak Dev University
Amritsar, India

Devershi Pallavi Bhatt
Manipal University
Jaipur, India

Soumi Bhattacharya
Narula Institute of Technology (NiT)
Affiliated under Maulana Abul Kalam
    Azad University of Technology
    (MAKAUT)
West Bengal, India

Manoj Kumar Bohra
Manipal University
Jaipur, India

Sandeep Chaurasia
Manipal University
Jaipur, India

Surbhi Chauhan
Jaipur Institute of Engineering
    Management
Jaipur, India

Nancy Girdhar
Amity University
Noida, India

Rakesh Gnanasekaran
Thiagarajar College
Tamil Nadu, India

Sarada Prasad Gochhayat
Virginia Modeling, Analysis, and
    Simulation Center, Old Dominion
    University
Suffolk, Virginia, USA

S. Gunasekaran
Ahalia School of Engineering and
    Technology
Palakkad, Kerala, India

Lokesh Gundaboina
Lovely Professional University
Punjab, India

Prashant Hemrajani
Manipal University
Jaipur, India

Abid Hussain
Career Point University
Kota, India

Vishal Jain
Sharda University
Greater Noida, U.P., India

Sandeep Joshi
Manipal University
Jaipur, India

Kailash Kumar
College of Computing and Informatics,
    Saudi Electronic University
Riyadh, Saudi Arabia

Chandrashekhar Kumbhar
Career Point University
Kota, India

Satpal Singh Kushwaha
Manipal University
Jaipur, India

Gaurav Malik
Ceridian
Toronto, Ontario, Canada

Kesana Mohana Lakshmi
CMR Technical Campus
Hyderabad, Telangana, India

Gnanasankaran Natarajan
Thiagarajar College
Tamil Nadu, India

D. Palanivel Rajan
CMR Engineering College
Hyderabad, India

Shobhandeb Paul
Guru Nanak Institute of Technology
    (GNIT)
Affiliated under Maulana Abul Kalam
    Azad University of Technology
    (MAKAUT)
West Bengal, India

Suneetha Rikhari
Mody University of Science and
    Technology
Lakshmangarh, Rajasthan, India

P. Rontala
School of Management IT and Governance,
    University of KwaZulu-Natal
Westville, Durban, South Africa

Siddhant Saxena
Manipal University
Jaipur, India

S. Shanmugam
Saveetha Engineering College
Chennai, India

Pulkit Sharma
Ceridian
Toronto, Ontario, Canada

Sandeep Sharma
Guru Nanak Dev University
Amritsar, India

Arjun Singh
Manipal University
Jaipur, India

Vineeta Soni
Manipal University
Jaipur, India

Manikumar Thangaraj
Thiagarajar College
Tamil Nadu, India

Narendra Singh Yadav
Manipal University
Jaipur, India

Introduction

Frontier technologies such as machine learning, blockchain, cyber security, virtual reality, and the Internet of Things have become integral components of our everyday activities. From artificial intelligence (AI) Cortana to machine learning (ML) Facebook face-detection algorithms, from user authentication in smart phones to investment in cryptocurrencies, all have paved the way for an innovative future. AI in its latest incarnation has developed from ML as a widely accepted application. ML first understands the structure of the data and then fits it into a statistical linear or non-linear model which can be used according to requirements. Blockchain represents an altogether new prospect for data administration and allows participants to work on live projects. There has been an unprecedented boom in cyber-attacks with consequent economic damage and loss of statistical data. Over the last decade, cyber security packages have increasingly come to depend on machine learning methods due to the remarkable scale of innovations in social networks, cloud technologies, e-commerce, the mobile environment, computer vison, IoT, and many other fields.

This book looks, among other applications pertaining to current technological developments, at the roles of AI in education, ML in analysing life-threatening diseases, cyber security for blockchain, and data security with blockchain. Our virtual tomorrow will have countless possibilities and currently we have seen but a small fragment of it. A digital ecosystem can give rise to major implications with a single technophile. The technological revolution will occur whenever it is economically flexible and a sufficient number of individuals have access to it. As these technologies emerge, so in this book we are looking for the challenges coming our way and providing appropriate solutions.

1

Intelligent Green Internet of Things: An Investigation

Salil Bharany and Sandeep Sharma

Guru Nanak Dev University, Amritsar, India

CONTENTS

DOI: 10.1201/9781003240310-1

1.1 Introduction

IoT research began in the 1980s, with the first IoT-connected device, a drink-vending machine, seen in 1982. 1989 saw the first 'future house' in the Netherlands, [1] targeting communication between man and machine. A toaster was connected to the internet in 1990 and modified by an inventor in 1991. The first machine-to-machine connection (MQTT) protocol, a standard publish-subscribe-based messaging protocol, was invented by IBM in 1999. In the same year, the Auto-ID Center was established at MIT [2] and related market analysis was published. LG promoted its smart refrigerator in 2000. It had an LCD screen displaying recipes, nutrition information, the freshness of stored foods, and temperature. The first IoT conference was held in 2008 in Zurich, Switzerland. Google launched the self-driving Google car concept in 2010. Innovation in IoT-assisted devices is ongoing.

The number of internet users, the number of devices, and the energy consumption levels have reached dangerous levels due to the expansion of the digital environment. By 2030, the number of associated devices worldwide will have reached 100 billion [3]. Scientists anticipate that the amount of content in 2030 will be10,000 times greater than in 2010, at the cost of a substantial rise in carbon emissions into the atmosphere, giving rise to environmental and health issues (Figure 1.1). By 2022, cellular networks will be emitting 445 million tonnes of carbon dioxide (CO_2), which is likely to rise [4, 5].

Current battery technology for devices has major implications for the future of green technology [6]. Experts estimate that 2022's 5G wireless communications will be capable of managing 1,000 times more mobile data than current cellular networks [7]. Using the TCP/IP protocol, the internet connects the world into a village of bits where everything is connected to the rest of the world.

The internet has had a profound impact on working life and social interactions [4]. A crucial aspect of the Internet of Things is the intelligent connection and context-aware computation of existing networks and system resources. With the IoT, everything is connected and communication anytime, anywhere, and of anything is enabled (Figure 1.1). IoT technologies are making machines more intelligent every day, allowing them to interpret data intelligently and communicate with each other more effectively. Devices such as radio frequency identification, sensors, actuators, mobile phones, and drones can communicate with each other and work together to achieve common goals. The many real-time monitoring applications include e-healthcare, home automation, environmental monitoring, and autonomous vehicles [8].

In the IoT, numerous intelligent agents are able to come to joint conclusions, and complete tasks in an optimal way [9]. The IoT is all about accumulating and analyzing data, and communicating with other devices in the rest of the world, as demonstrated in Figure 1.2.

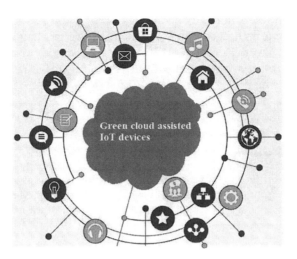

FIGURE 1.1
Green IoT environment.

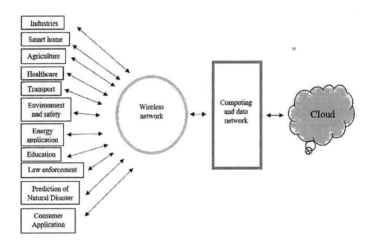

FIGURE 1.2
Basic working of IoT.

ICT technologies can contribute to global climate change [2] since more and more energy is utilized by ICT applications. Big data requires a lot of storage space, cloud computing, and bandwidth. The considerable power needed to analyze large amounts of data will increase pressure on society and the environment, and green IoT aims to minimize carbon emissions and energy usage. Technologies that have built the eco-friendly IoT enable their users to acquire, store, access, and handle a broad range of data in an energy-efficient way; these are referred to as green ICT technologies [10]. The optimization of data centres by spreading the infrastructure increases energy efficiency, and reduces CO_2 emissions and the disposal of e-waste [4]. Green ICT technologies bring numerous advantages to society, such as reducing the energy needed to manufacture ICT devices, including the production and distribution of machines and equipment.

With green IoT, computers and servers are designed, manufactured and disposed of in a way that will not harm the environment, increasing efficiency and decreasing their impact on society and the environment [1]. To decrease the IoT's impact on human health and environment is our key priority [11], so new resources are being sought, with the. primary objective of reducing power usage, carbon emissions, and pollution by taking advantage of ecological conservation [12].

Green IoT has significant potential to assist both the economic process and environmental sustainability with green ICT technologies [13]. With the help of these intelligent and developing technologies, the world is becoming eco-friendlier and more intelligent. However, a critical assessment of methods and schemes for green IoT is lacking, so this paper examines the fundamentals of green IoT technologies, reviewing previous work on building a green and intelligent world. The rest of the chapter is organized as follows. An overview of green IoT is provided in Section 1.2. Section 1.3 covers relevant surveys. Section 1.4 describes the layered approach to IoT. Sections 1.5 and 1.6 discuss IoT applications and protocols respectively. Issues and challenges in IoT are described in Sections 1.7 and 1.8. Section 1.9 examines future directions for green IoT, with future green IoT work and closing remarks in Section 1.9.1.

1.2 Green IoT

IoT architecture is building a green environment by leveraging the most modern technology smartly and effectively, using cameras and intelligent sensors to build a global, inconspicuous, immersive close communication network and coding environment [14]. To achieve intelligence and sustainability and reduce CO_2 emissions, green IoT concentrates on cutting down IoT energy consumption. The authors of [13] outline several different technical approaches to easily achieving a greener internet. Uddin et al. [15] have provided strategies for improving energy efficiency and lowering CO_2 by shifting to green information technology [16]. Green IoT plays a vital role in the dissemination of IoT and the reduction of CO_2 emissions [15], using environmental conservation ideas [17], and cutting down power consumption [16] to achieve an innovative and sustainable society.

According to Murugesan, green technology is "the study and practice of developing, utilizing, producing, dismantling and disposing of servers (computers), monitors (storage devices), printers, and communication network systems with minimum or no environmental impact".[18]. Examples of design technologies include the power consumption of devices as well as communication methods, network topologies, and connectivity technologies. Green IoT technologies are significantly impacting the minds of users, contributing to preserving natural resources, reducing costs, and reducing the impact of technology on the environment and human health. Green IoT focuses on green uses, green utilization, green planning, and green manufacturing [19].

1. *Green uses*: employing computers and other ecologically friendly information systems while limiting their power consumption.

2. *Green utilization*: recycling old computers and other electronic equipment by repairing and repurposing outdated computers.

3. *Green planning*: construction of green IoT components, computers, and cooling equipment that must be energy efficient and environmentally friendly.

4. *Green manufacturing*: environmentally friendly production of electronic components, computers, and other subsystems.

1.3 Related Surveys

A detailed assessment of energy conservation challenges and solutions was conducted by Abbas et al. [20]. The study examined different operational characteristics of IoT devices using 3GPP machine communication, including IEEE-802.11ah, Bluetooth, and Z-wave, congestion avoidance methods, and heterogeneous radio interfaces. The authors also suggested future areas for study relating to energy conservation and efficiency challenges. Marcus et al. [21] described a novel MAC protocol for wireless sensor networks, called PaderMAC, which uses TinyOS and MAC layer architecture. The goal was to cut down the energy consumption to increase the lifespan of IoT devices. A group of researchers led by Algimantas et al. [22] developed an energy-efficient SSL protocol, ensuring high bandwidth and the necessary degree of security with minimal energy usage. In addition to explaining the SSL protocol's fundamentals, they also suggested an adaptive SSL protocol that incorporated confidentiality, integrity, and accessibility. Cryptography proficiencies were used to accomplish SSL protocol security goals. In their study, Chen et al. [23] summarized topology-based routing protocol and data link protocol. Having a solid topological structure can improve routing protocol performance, extending the network lifespan by providing the basis for data fusion and objective placement features. The authors explore WSN energy supply and management strategies. Kim et al. [24] proposed an energy-efficient and reliable data automated repeat request method comprising duplication re-transmission prevention, congestion control, and error reporting. Talwar [25] described routing strategies and protocols for IoT, such as RPL, OLSR, and PRoPHET, including some of the main difficulties.

Energy-efficient approaches for both the physical layer and deployment are presented by Wu et al. [26]. They also introduced an energy-efficient optimization principle. There are many energy-related algorithms, such as multi-level water filling or bi-section algorithms for improving energy use. Suo et al. [27] studied encryption mechanisms, communication security, and sensor data protection. They also looked at security features and needs from four different layers: the perception and network layers, and the support and application layers. Beaudaux [28] proposed a heterogeneous MAC duty-cycle configuration among nodes in the network. The nodes are divided into distinct groups, each assigned a specific duty-cycle design. Routing role and sleep depth are discussed in the suggested approach. It was formerly possible for MAC and routing layers to work together, such that each router knew how many nodes were above it in the routing tree. As a result, the nodes were divided into distinct sleep-depths, represented as discontinuous virtual layers, based on applicative criteria. Using an activity scheduling method for sensing coverage, Lu et al. [29] proposed a technique for communicating activity decisions in which nodes listen to messages from neighbours and must wait until the timeout has elapsed for each round before selecting a timeout.

Abdullah et al. [30] presented an architecture using a cluster-based routing method and a message scheduling technique in which sensors are grouped to form a single unit. The IoT subgroup is a collection of sensor nodes where any node can act as a base at any time. The base node aggregates the data from neighbouring nodes and delivers it to the base station for transmission. Sensor messages are analyzed and messages reorganized based on queuing theory. A detailed review of energy-efficient routing protocols in WSN is given by Singh et al. [31]. Energy-efficient dependable barriers are examined by Sundar et al. [32], who provide instances and recommend a scheduling method as a solution. In this plan, time slots are necessary, and activities are scheduled in rounds. By broadcasting a message, the node builds its own activity decision and communicates it to its neighbours. Due to the limited number of control messages sent, this method is energy efficient.

A cluster-based sleep scheduler for M2M communication networks was proposed by Mohammed et al. [33]. In this system devices were positioned, and clusters were established. Initially, all instruments were considered to have the same amount of power (energy). A plethora of devices was chosen to be the Principal Cluster Head (PCH). Many devices were developed to act as alternative cluster heads (ACHs) for PCH devices. The PCH determined the number of active devices in each cluster. All of these devices were able to offer network coverage and were always on. Inactive/sleep mode was maintained on all other devices. As a result, energy usage was reduced compared to other techniques. As part of a study on energy efficiency routing methods in wireless sensor networks, Katkar et al. [34] discussed how to make them more fault tolerant [16] while still maintaining accuracy and scalability to deploy nodes and reduce energy usage. Table 1.1 shows all the studied techniques.

TABLE 1.1

Issue and Challenges in All Related Surveys

References	Challenges	Description of Technique	Category
Abbas et al. [20]	Energy efficiency	Duty-cycle, overload and link radio resources	Software
Marcus et al. [21]	Sleep and wake-up periods	Pader MAC protocol	Software
Algimantas [22]	Encryption and authentication	SSL protocol maximum bandwidth	Software
Chen et al. [23]	Energy in WSN	Protocols for routing and topology control	Hardware
Kim [24]	Power consumption	ARQ request scheme	Software
Talwar [25]	Latency, throughput, and node energy consumption	Improved routing protocol	Software
Wu et al. [26]	Efficiency in spectrum usage	Optimization constraints, variables and algorithms	Hardware
Suo [27]	Privacy and security	Encryption, communication, and cryptographic algorithms	Software
BeaudauxE [28]	Inefficient energy usage	Heterogeneous duty cycle of MAC	Software
Rongxing Lu et al. [29]	Security and reliability	Energy-efficient scheme	Software
Abdullah et al. [30]	High service time and inefficient energy usage	Scheduling scheme for messaging	Hardware
Singh et al. [31]	Inefficient energy consumption	Efficient routing scheme	Hardware
Sundar [32]	Inefficient energy usage and low reliability	Optimized scheme for scheduling	Software
Mohammed et al. [33].	Node-sleeping scheduling schemes	Energy-efficient mobility-centric node scheduling scheme based on clustering	Software
Katkar et al. [34]	Inefficient routing	Survey on routing protocols	Software

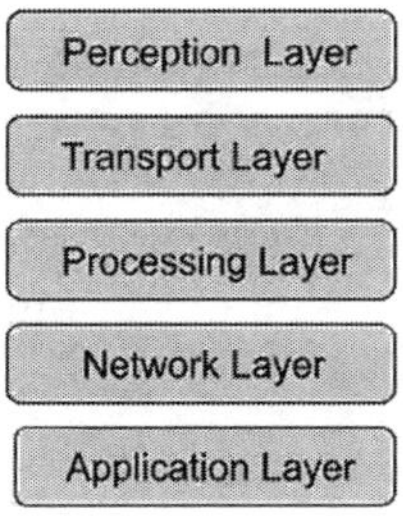

FIGURE 1.3
Layers of IoT.

1.4 IoT Layered Architecture

This section describes the power consumption components associated with each layer. A layered architecture is used to collect and analyze a massive amount of data [35]. Hardware and software techs, protocols for communication, and various processing technologies combine to form this hybrid system. Applications, QoS considerations, interoperability and other factors influence IoT architectures. They also focus on the most power-hungry nodes and protocols, as well as middleware parts and application programs. Perception, transport, processing, network, and application are the five levels of the architecture (Figure 1.3).

1.4.1 Perception Layer

This layer collects data, schedules tasks, and communicates with other sensors. Self-coordinated sensing and load balancing are two of this layer's primary functions. Among the components of this layer are passive radio frequency identification (RFID), sensors, sink nodes, actuators, gateway nodes, and software results (such as bootstrap software), all of which are high in energy consumption [36]. Sensor examples include medical, military, chemical, ADC, accelerometers, cameras, GPS, etc.

1.4.2 Transport Layer

Perception is communicated to processing through this layer, the components of which include 3G, LAN, Bluetooth and RFID, NFC, WIFI, coordinators and trust centre gateways, and master–slave Bluetooth [5]. Tags, readers, electrometers, short-range communicating protocols, trust centres, and gateways are examples of transport layer components [35].

1.4.3 Processing/Middleware Layer

A vast amount of data is dissected, treated, and stored in this layer of the system, which supplies services to lower levels, apportions resources for efficient storage in virtual and real computers, and transforms data into the appropriate format for use [37]. Middleware and databases are also part of this layer. Servers, data centres and microcontrollers are examples of processing elements. Middleware includes Contiki, TinyOS, RTOS, and databases such as data warehouses and allocator CPUs, and semantic and service-based middleware.

1.4.4 Network Layer

As a total system model, this layer facilitates communication among devices at other levels. It includes WSN nodes, cloud hosts, and large data gateway nodes, all of which consume a lot of electricity [37]. Hubs and switches are examples of WSN elements, such as low-power devices, sensor and processing elements, power and routing features, IoT cellular carriers, and distant transmitting nodes.

1.4.5 Application Layer

This is an abstraction layer that allows consumers to access services from the underlying levels. It visualizes processed data as intelligent [38] activities. There are several examples of applications in which control systems consume a lot of power, including intelligent homes, intelligent buildings, smart cars, healthcare, intelligent environments, supply-chain management, energy conservation, etc.

1.5 Applications

1.5.1 IoT in Industry

IoT provides a digitally connected factory which monitors production flow, including inventory management, plant safety and security [6], quality control and environmental gases like oxygen and ozone. Asset management or asset tracking are carried out with the help of Bluetooth, Wi-Fi, or LoRaWAN devices.

1.5.2 IoT for the Smart Home

IoT is involved in aspects of home automation such as lighting control, home safety, and security. It includes smart switches, HVAC, landscape maintenance, device monitoring, air and water quality, and customized delivery of infotainment [1]. Artificial intelligence voice assistance and digital experiences, and smart locks, are also part of this application. Smart energy meters [39] use various IoT components like gateways, sensors, firmware, cloud and databases, protocols, and middleware. A smartphone, computer, and tablet can control a modern smart home.

1.5.3 IoT for Agriculture

IoT in agriculture can help farmers reduce waste and enhance productivity. IoT devices are used in precision farming [2] to control crop growth. Various components like detectors, sensors, autonomous vehicles, automated hardware, control systems, and robotics are used. Drones are used for crop assessment, irrigation, planting, soil and field analysis. IoT is also used in livestock management and intelligent greenhouses.

1.5.4 IoT in Healthcare

IoT can monitor and check the progress of health conditions using wearable devices like hearing enhancers, and mood sensors to improve quality of life [40]. The IoT has applications in diseases like diabetes, Alzheimer's, cardiac arrhythmia, etc.

1.5.5 IoT in Transport

There is significant AI-based development in transport services. The merger of various autonomous processes such as information processing, control, decision-making, and communication with other devices can be done by IoT devices for services such as automatic traffic management [39], automatic parking, electronic toll payment systems, road safety and roadside assistance. In future, we will also be able to minimize pollution and maximize efficient mobility. We can significantly reduce traffic accidents and save lives [41]. Autonomous vehicles will help people with disabilities and those who are unable to drive to increase their independence [17].

1.5.6 IoT in Environment and Safety

IoT is involved in town and city planning, defence, management of the economy, and law enforcement. Relevant aspects include population growth, zoning, mapping, water and food supply, social services, land use, and transportation patterns. IoT collects data, analyses it and gives a more accurate output [9].

1.5.7 IoT in Energy Applications

IoT provides various monitoring and energy control functions for multiple devices and energy sources with commercial and residential applications [11]. Using IoT devices, we can quickly locate and fix leaks or damage in machines or supply lines.

1.5.8 IoT in Education

This technology allows the quality of education, professional development, cost control, and facilities management to be improved.

1.5.9 IoT in Law Enforcement

IoT is used in policing and court systems [40]. In a court system, IoT improves satisfaction, reduces costs, manages corruption, eliminates excessive court procedures, enables storage of evidence, and provides superior analytics.

1.5.10 IoT in the Prediction of Natural Disasters

Sensors and their self-governing coordination and simulation will enable prediction of natural disasters [11] so that proper steps can be taken in advance.

1.5.11 IoT in Consumer Applications

IoT technology enhances the human way of life, both at work and at play. IoT acts like an adviser and personal assistant, and provides security: it can play the part of butler, gardener, security guard, chef, or odd-job man. At the workplace, IoT follows your work routine, adjusts the surrounding environment, and acts as a consultant and manager, increasing output and reducing working time. It provides professional support, as mentioned by the manufacturer.

1.6 IoT Protocols

The many IoT communication protocols can be classified in two groups [42]:

1. Low energy WAN – Sigfox, cellular
2. Short-range network – Bluetooth, ZigBee, RFID, NFC, Z-wave, 6lowPAN.

Any protocol can be used according to its characteristics: frequency band, topology, network, power, data rate, security, range, modulation type [43].

1.7 Limitations and Future Research Directions

Despite massive research efforts, green IoT technology is still in its initial phase and numerous problems remain to be resolved. Among the most pressing issues are:

- Efficient security mechanisms.
- Security is a critical issue in industrial applications. Data hacking and the absence of industry standards are the major challenges. Although there are many protocols for manufacturing and industrial settings, there is no standardization to ensure interoperability.
- Many IoT devices do not support any modifications or enhancements. Most of the applications are on the IoT. There may be big data with high data complexity.
- Transparency regarding functionality is often lacking in IoT devices, and there is no control over unwanted functions or data collection.
- The unpredictable behaviour of IoT devices means any system is well designed, defined and within governing body control. However, there can be no certainty about how it interacts with other devices.
- Achieving satisfactory performance by integrating energy efficiency across the IoT architecture.
- Apps ought to be eco-friendly to reduce their impact on the planet.

1.8 Issues in Energy Conservation

According to the literature, energy efficiency in the IoT can be summarized as follows.

1.8.1 Idle Listening

An active node is a significant energy consumer. It is crucial to reduce the amount of energy wasted. It does not have to be in a state (active). Idle listening is the state of being ready to communicate data while not receiving or transmitting packets. There are various ways to minimize the total amount of time spent active. After a set time or when a wake-up signal is processed, sleeping sensor nodes are reactivated.

1.8.2 Collision

Nodes collide when they receive several data packets at once, rendering the data obtained worthless. While energy is dissipating, the transmission process must be repeated. Collisions also increase the latency of the system [44]. All of these transactions may use a much higher amount of energy.

1.8.3 Overhearing

Due to the high density of sensor nodes, data transmission is hampered by interference from neighbouring nodes. The technical term for this is overhearing. This is an issue that only affects the nodes within reach. Much energy is wasted receiving and processing meaningless data [45].

1.8.4 Reduction of Protocol Overheads

The protocol header entropy uses high levels of energy. On-the-fly transmission intervals and cross-layering are among the ways to reduce protocol overheads.

1.8.5 Traffic Fluctuations

As a result of heavy traffic, there may be lengthy delays. At maximum capacity, congestion can reach a dangerously high level.

1.9 Energy Preservation Approaches

Analysis of the literature suggests several ways to conserve energy in IoT.

1.9.1 Node Activity Management

In node activity, there are two components: sleeping scheduling and on-demand node activity. When a node is placed in sleep mode, a sleep schedule is used to decide when to wake up. This conserves energy by reducing the amount of time spent inactive. There are specific times when each node is in a dormant state. Nodes in the region that receive a wake-up signal go into active mode. When the device is activated, data is sent. A sensing coverage and an activity scheduling method [13] are also proposed. There are several rounds of this exercise. When a node starts a new process, it chooses a random timeout and listens to messages from its neighbours until it chokes before continuing. The communications contain a conclusion to be active or not, which the recipient may make. This form of communication requires only a tiny amount of energy. Increasing energy efficiency through lowering energy usage at the node level is a possibility.

1.9.2 Data Aggregation and Transmission Process

Data transmission is more expensive than data processing. Aggregating data inside clusters may be quite helpful. Because cluster heads are responsible for monitoring and processing requests, groups can minimize the quantity of data, reducing the consequences of energy dissipation in a variety of ways. Data originating from many sources are integrated into one packet, reducing duplication and communications. Wireless data transmission consumes a substantial part of the entire amount of energy available to us. Increased energy savings may result from including power control in the transmission process [46]. Transmission power, which is associated with data rate, and circuit power should be cautiously equilibrated in short-range applications to achieve high energy efficiency. Optimization techniques are employed to achieve energy efficiency during transmission.

1.9.3 MAC Protocol

IoT devices require a great deal of energy to operate. One approach is to conserve energy by improving the MAC protocol.

In the data connection layer, the MAC protocol is regarded as a sublayer. It establishes the rules for transmitting the frame in a certain way. It is the MAC protocol that organizes channel access when there are many nodes. The IEEE published a popular MAC standard in 2003, updated in 2006. Non-beacon mode and beacon mode are defined by IEEE 802.15.4. To obtain a frame, the former must be awake and alert at all times. The latter describes a superframe in which nodes are only asleep for part of the frame period: energy usage and throughput rise. By placing the node into a sleep state, the duty-cycle protocol aims to eliminate unnecessary activity. A periodic wake-up method is used for this. Periodically, a node wakes up to send or receive packets, and if there is no activity nodes go to sleep. As a result, the duty cycle is calculated as the ratio of listening time to wake-up time. As synchronous and asynchronous protocols, low duty-cycle protocols come in a variety of forms. In computing, the idea of synchrony refers to the flow of data between computers. When it comes to asynchronous communication, there are two main concepts: transmitter initiated and receiver initiated. In the first method, nodes send request packets until they reach their destination. In the second technique, a node sends out packets to neighbouring nodes to tell them that it is willing to accept them. For low-power operation, a variety of MAC protocols have been developed. MQTT, XMPP, DDS, and AMQP are examples of IoT protocols. Encryption of data is therefore required to ensure its security. The more stringent the safety precautions, the more resources are used. This makes it impossible to use current communication security methods, which are more energy intensive, at the network layer. When it comes to this layer, it is vital to maintain confidentiality and integrity. In the application layer, authentication and key agreement over a diverse network are two important factors to consider. The systems are not built for devices that have a limited amount of resource. It is, therefore, necessary to construct lightweight cryptographic methods in this layer.

1.9.4 Security Management

Node security mechanisms must take energy into account. However, a resource-constrained gadget does not lend itself to security solutions that are built for it. Making encryption methods quicker and less energy consuming is one of the problems. To increase battery life, it is necessary to reduce energy usage. Energy consumption of devices is significantly affected by security encryption and decryption functions. Each IoT layer has its own security requirements. Authentication is required at the perceptual layer to prevent unwanted node access, so encryption of data is necessary to maintain its confidentiality. Enhanced safety procedures require additional resources. There are currently no energy-efficient communication security techniques that can be used at the network layer. Confidentiality and integrity are essential in this tier. Lightweight cryptographic techniques must be implemented at the application layer, and authentication and critical arrangement across a heterogeneous network are important aspects to be addressed. Systems are not designed to work with devices with restricted resources.

1.9.5 Routing

As the name implies, routing is moving information from a source to a destination through a network. To route something, you have to decide on which paths to take. Scheduling may be broken down into three types: flat-based, hierarchical, and location-based routing. There are no roles of hierarchy functions in flat-based routing. Routing nodes will have distinct responsibilities in hierarchical routing. Localized routing uses node locations to send data. Routing protocols include multi-path, query, negotiation, QoS, and coherence. We may classify routing protocols as either proactive, reactive, or mixed. To keep track of the network's topology, proactive protocols seek to acquire routing information proactively. In a reactive protocol, routes are found on demand by searching for them [35]. The route-finding procedure is only started when a transmission begins. In mixed protocols, a route's complete path is used for data transfer. As a result, memory use is minimized [46] but the size of headers and the amount of traffic is increased. The multipath-routing protocol searches for different routes to reach any destination. Hop count is the most often used metric. The route with the smallest number of hops is selected. Node-level or network-level energy metrics can be used to affect protocol routing decisions so that energy resources are conserved [39].

1.10 Conclusion

In this chapter, we examined the green IoT perspective and highlighted key IoT challenges, security issues, and safety countermeasures from the energy-saving perspective. We talked about current initiatives in the field of green IoT, and highlighted possible applications for green IoT employed within clouds in the future. IoT apps that conserve energy with green settings were listed and future recommendations for enhancing green IoT applications were discussed. We also highlighted the need for standards in green design to maintain consistency across several areas of green IoT, mostly in connection with heterogeneous communication, integration with sensor cloud, and the provision of highly reliable service management in complex physical conditions.

References

1. S. Murugesan, Harnessing green IT: Principles and practices, *IT Professional*, 10 (2008) 24–33.
2. A. Mickoleit, *Greener and smarter: ICTs, the environment and climate change*, In OECD Green Growth Papers. OECD Publishing (2010). https://doi.org/10.1787/5k9h3635kdbt-en

3. Accenture Strategy, SMARTer2030: ICT solutions for 21*st* century challenges, *Global eSustainability Initiative (GeSI)*, Brussels, Belgium, Technical Report (2015).

4. L. Atzori, A. Iera, G. Morabito, The internet of things: A survey, *Computer Networks*, 54 (2010) 2787–2805.

5. Green Power for Mobile, The Global Telecom Tower ESCO Market, Technical Report (2015).

6. IMT Vision-Framework and Overall Objectives of the Future Development of IMT for 2020 and Beyond, document Rec. ITU-R M.2083- 0 (2015).

7. M. Albreem, 5G Wireless communication systems: vision and challenges, *2015 IEEE International Conference on Computer, Communication, and Control Technology*, Malaysia (2015).

8. D. Popa, D.D. Popa, M.-M. Codescu, Reliability for a green internet of things, *Buletinul AGIR nr* (2017) 45–50.

9. S.S. Prasad, C. Kumar, A green and reliable internet of things, *Communications and Network*, 5 (2013) 44.

10. C. Zhu, V.C. Leung, L. Shu, E.C.-H. Ngai, Green internet of things for the smart world, *IEEE Access*, 3 (2015) 2151–2162.

11. Y.-L. Lü, J. Geng, G.-Z. He, Industrial transformation and green production to reduce environmental emissions: Taking cement industry as a case, *Advances in Climate Change Research*, 6 (2015) 202–209.

12. A. Gapchup, A. Wani, A. Wadghule, S. Jadhav, Emerging trends of green IoT for smart world, *International Journal of Innovative Research in Computer and Communication Engineering*, 5 (2017) 2139–2148.

13. R. Arshad, S. Zahoor, M.A. Shah, A. Wahid, H. Yu, Green IoT: An investigation on energy saving practices for 2020 and beyond, *IEEE Access*, 5 (2017) 15667–15681.

14. Danita, M., Mathew, B., Shereen, N., Sharon, N., Paul, J. J. (2018). IoT based automated greenhouse monitoring system, In *2018 Second International Conference on Intelligent Computing and Control Systems (ICICCS)*, IEEE. https://doi.org/10.1109/iccons.2018.8662911

15. M. Uddin, A.A. Rahman, Energy efficiency and low carbon enabler green IT framework for data centers considering green metrics, *Renewable and Sustainable Energy Reviews*, 16 (2012) 4078–4094.

16. B. Talwar, S. Bharany, A. Arora, Proactive detection of deteriorating node based migration for energy-aware fault tolerance, *Think India Journal*, 22 (2019) 25.

17. Zima, H. P., Nikora, A., Fault tolerance. In *Encyclopedia of parallel computing*, ed. David Padua (2011) (pp. 645–658). Springer US. doi.10.1007/978-0-387-09766-4_63

18. S. Murugesan, G. Gangadharan, *Harnessing green IT: Principles and practices*, Wiley Publishing (2012).

19. C.S. Nandyala, H.-K. Kim, Green IoT Agriculture and Healthcare Application (GAHA), *International Journal of Smart Home*, 10 (2016) 289–300.

20. Zeehan Abbas, Wonyong Yoon, A survey on energy conserving mechanism for the internet of things: Wireless networking aspects, *Sensors*, 15 (2015) 24818–24847.

21. Marcus Autenrieth, Hannes Frey, PaderMAC: Energy-efficient machine to machine communication for cyberphysical systems, *Peer-to-Peer Networking and Applications*, 7 (2014) 243–254.

22. Algimantas Venckauskas, Nerijus Jusas, Egidijus Kazanavicius, Vytautas Stuikys, An energy efficient protocol for the internet of things, *Journal of Electrical Engineering*, 66 1 (2015) 47–52.

23. Fangxin Chen, Lejiang Guo, Chang Chen, A survey on energy management in the wireless sensor networks, *IERI Procedia*, 3 (2012) 60–66.

24. Kyungmin Kim, Jaeho Lee, Jaiyong Lee, Energy efficient and reliable ARQ scheme (E2 R-ACK) for mission critical M2M/IoT services, *Wirless Personal Communication*, 78 (2014) 1917–1933.

25. Mallikarjun Talwar, Routing techniques and protocols for internet of things: A survey, *Proceeding of NCRIET & Indian Journal of Scientific Research*, 12 (2015) 417–423.

26. Gang Wu, Chenyang Yang, Shaoqian Li, Geoffrey Ye Li, Recent advances in energy-efficient networks and their applications in 5G systems, *IEEE Wireless Communication*, 222 (2015) 145–151.

27. Hui Suo, Jiafu Wan, Security in the internet of things: A review, *IEEE*, 3 (2012) 648–651.

28. Julien Beaudaux, Antoine Gallais, Thomas Noel, *Heterogeneous MAC duty-cycling for energy-efficient internet of things deployments*, Tsinghu University Press and Springer-Verlag Berlin Heidelberg, 3 1 (2013), 54–62.

29. Rongxing Lu, Xu Li, Xiaohui Liang, Xuemin Shen, GRS: The green, reliability, and security of emerging machine to machine communications, *IEEE*, 49 4 (2011) 28–35.

30. Saima Abdullah, Kun Yang, An energy-efficient message scheduling algorithm in internet of thing environment, *IEEE*, 2013 311–316.

31. Satvir Singh, A survey on energy efficient routing in wireless sensor networks, *International Journal of Advanced Research in Computer Science and Software Engineering*, 3 7 (2013) 184–189.

32. Shyam Sundar Prasad, Chanakya Kumar, A green and reliable internet of things, *Communications and Networks*, 5 (February 2013) 44–48.

33. Mohammed S Al-Kahtani, "Efficient cluster-based sleep scheduling for M2M communication network", *Arabian Journal for Science and Engineering*, 40 8 (2015) 2361–2373.

34. Pallavi S Katkar, Vijay R Ghorpade, A survey on energy efficient routing protocol for wireless sensor networks, *International Journal of Computer Science and Information Technologies (IJCSIT)*, 6 1 (2015) 81–83.

35. Andrew Whitmore, Anurag Agarwal, Li Da Xu, The internet of things—A survey of topics and trends, *Information Systems Frontiers*, 17 2 (2015) 261–274.

36. Qiu, T., Zhao, A., Ma, R., Chang, V., Liu, F., & Fu, Z., A task-efficient sink node based on embedded multi-core SoC for Internet of Things. *Future Generation Computer Systems, 82* (2018), 656–666. https://doi.org/10.1016/j.future.2016.12.024

37. Chunsheng, Zhu, et al., Green internet of things for smart world, *IEEE Access, 3* (2015) 2151–2162.

38. Keyur K. Patil, Sunil M. Patel, Internet of Things-IOT: Definition, characteristics, architecture, enabling technologies, application & future challenges, *International Journal of Engineering Science and Computing*, 6 5 (2016) 6122–6131.

39. S. Bharany, S. Sharma, S. Badotra, OI. Khalaf, Y. Alotaibi, S. Alghamdi, F. Alassery, Energy-efficient clustering scheme for flying ad-hoc networks using an optimized LEACH protocol, *Energies*, 14 19 (2021) 6016. doi.10.3390/en14196016

40. H.-I. Wang, Constructing the green campus within the internet of things architecture, *International Journal of Distributed Sensor Networks*, 10 (2014) 804627.

41. P. Kumari, P. Kaur A survey of fault tolerance in cloud computing. *Journal of King Saud-University-Computer and Information Sciences*, 33 (2018), 1159–1176. doi.10.1016/j.jksuci.2018.09.021

42. A.L. Di Salvo, F. Agostinho, C.M. Almeida, B.F. Giannetti, Can cloud computing be labeled as "green"? Insights under an environmental accounting perspective, *Renewable and Sustainable Energy Reviews*, 69 (2017) 514–526.

43. J. Gubbi, R. Buyya, S. Marusic, M. Palaniswami, Internet of Things (IoT): A vision, architectural elements, and future directions, *Future Generation Computer Systems*, 29(2013) 1645–1660.

44. F.K. Shaikh, S. Zeadally, E. Exposito, Enabling technologies for green internet of things, *IEEE Systems Journal*, 11 (2015), 983–994.

45. C. Xiaojun, L. Xianpeng, X. Peng, IOT-based air pollution monitoring and forecasting system, *Computer and Computational Sciences (ICCCS), 2015 International Conference on*, Greater Noida, India, IEEE (2015), pp. 257–260.

46. S. Manna, S.S. Bhunia, N. Mukherjee, Vehicular pollution monitoring using IoT, *Recent Advances and Innovations in Engineering (ICRAIE)*, 2014, IEEE (2014), pp. 1–5.

2

The Role of Artificial Intelligence in the Education Sector: Possibilities and Challenges

Riya Aggarwal and Nancy Girdhar

Amity University, Noida, India

Alpana

Manav Rachna University, Faridabad, India

CONTENTS

2.1 Introduction

With the rise of computing resources, expanded refinement of algorithms, and the mass of information, AI is venturing into numerous spaces of everyday life. Artificial intelligence is a far-reaching subsection of computer science referring to the integration of human intelligence with robots. "AI is a new energy", just as electricity fueled every part of human activities a decade or a century ago [1]. From Siri to stair-climbing wheelchairs, from CCTV cameras to chatbots, from automatic parking systems to intelligent tutoring systems (ITSs), from mobile phones to touchscreen laptops, from projectors in classrooms to students using tablets in the classroom, from normal robots to computer vision-based self-drawing robots, from academic counselling to curriculum to personalized learning—artificial

DOI: 10.1201/9781003240310-2

intelligence is growing ever more rapidly. A survey shows that in about 60% of jobs, at least 33% of the tasks could be completely automated, suggesting considerable changes in the work environment and culture. In the context of education, acceptance of AI is also moving fast. More than 50% of today's students will be engaged in jobs that do not exist yet, as AI assumes control over jobs generally taken by people [2]. This has prompted worries from teachers and government authorities about how much longer people will have a place in this world. According to a recent survey by IBM,[1] forums such as Burning Glass Technologies and the Business-Higher Education Forum predict that the number of jobs in data analytics in the USA will rise by 364,000 to 2,720,000 by 2022. Researchers and inventors are making astonishingly rapid improvements in stimulating activities such as learning, reasoning, and assessment, to the extent that these can be accurately defined. Before long designers may have the option to foster frameworks that surpass the limits of human learning or reasoning abilities. As innovation progresses, benchmarks that had previously characterized AI become obsolete. For instance, devices that figure fundamental capacities or perceive text through optical recognition are not, at this point considered to typify AI, since this capacity is presently underestimated as an intrinsic PC work. AI is delivering revolutionary changes at the centre of numerous industry verticals, and it is being investigated in the education sector as well [3].

AI is not only changing educational programmes in science, innovation, arithmetic, and engineering, but is reshaping the whole education system. With widespread applications in customer service, assembly, and quality control, AI is poised as the main innovation for future advances. From anticipating traffic designs to suggesting items to be purchased, to recognizing great "matches" between people, individuals are becoming increasingly dependent on algorithms to help them make choices in their everyday lives [4].

This chapter presents the key challenges faced by the modern education system, and the AI-based solutions that have been and can be used to overcome them, with a focus on master systems. Enhanced AI applications for learning during the current Covid-19 crisis are analyzed. AI software advances for students and teachers currently available on the market are reviewed and assessed. A detailed literature survey evaluating publications on AI and education between 2013 and 2021 is conducted [5]. We compare the role of big-tech companies that are making AI the game-changer in the educational sector. The chapter also considers the current advancement of AI in the real world. The study seeks to establish the impact of fourth industrial revolution (IR 4.0) items including AI, the internet of things, natural language processing (NLP), cloud computing, cybersecurity, big data, and blockchain on learning analytics and academic achievements, and their social implications [6]. A cross-disciplinary thematic analysis of research papers published in prestigious journals has been conducted.

The chapter is organized as follows: A detailed backdrop to the research is provided in Section 2.2. The literature review follows in Section 2.3. The findings of the research survey are discussed in Section 2.4, along with the proposed framework. Section 2.6 concludes the work with some future research directions indicated in Section 2.6.

2.2 Background to the Study

AI has great potential in education to produce personalized learning systems, dynamic evaluation, and online learning experiences. The computing and data filtering techniques that have brought about "AI + Education" include intelligent tutoring systems, adaptive learning, learning data analytics, human–robot interactions, educational software like Grammarly, and so on. Customized adaptive AI learning technologies adapt to the learner's learning techniques, task sequences and complexity, reporting time, and choices [7]. Temporary disruptions to traditional classrooms and learning during the Covid-19 pandemic shifted education from an offline to online paradigm. The education landscape is evolving rapidly, and educational technology has a significant role to play as establishments adjust to massive changes in students' inclinations and assumptions. It is widely acknowledged that in filling gaps in learning and instruction, AI is empowering education to make greater strides than ever before [8].

2.2.1 History of AI

Artificial intelligence was first officially recognized in 1956 at a conference held at Dartmouth College in the United States [9]. The history of AI consists of work and experimentation not just by mathematicians and computer analysts but also by therapists, physicists, psychologists, and market analysts. The chronology spans the pre-1950 era of statistical techniques to the present day, including Alpha Zero in 2017. The most significant advance in the development of technology occurred during WWII, when both the Allied powers and their adversaries sought to foster innovation to assist them to victory [9]. As a result, a substantial amount of money is available for research and development at numerous universities. Artificial intelligence advancements have provided the world with computers that can defeat humans at chess and *Jeopardy!*, as well as drive vehicles and manage hectic schedules. Nevertheless, technologists are still decades away from creating self-aware computers—seen by some as a panacea to solve poverty and disease, and feared by others as putting humanity's survival at risk [10].

Artificial intelligence has evolved from the depths of academic research to the foreground of public debate, including at the United Nations, in only five years, owing to some notable accomplishments and to its transformative potential [10].

2.2.2 AI in Education

The use of AI technology or its application in academic contexts to assist education, training, or assessment is referred to as AIED (Artificial Intelligence in Education). The more profound the influence of AI on mankind, the more urgent it is for us to comprehend it. The establishment of the Beijing Consensus (UNESCO, 2019) compelled officials and educators worldwide to perceive the importance of outlining the future direction of education [11]. The adoption of AI in education is being hailed as a game-changer, with students acquiring more knowledge than they could ever have received from a single teacher. AIED, a research field established almost 30 years ago, has emerged as a critical strategic goal in several countries and regions, notably in K-12 education. Implementing AIED has opened up new possibilities for creating effective teaching strategies and improved AI-based learning applications or systems [12]. Technical approaches that incorporate both qualitative and quantitative aspects can be used to forecast a range of scenarios from idealistic objectives at one end to cautionary tales at the other. As progress in the fields of processing and neuroscience continue to reduce the gap between science fiction and reality, it is useful to pause for a minute to perceive how we got to where we are today with AI and identify the critical moments along the journey for growing genuine machine knowledge.

TABLE 2.1

Brief History of AI [9]

Year	Description
ALAN TURING—TEST FOR MACHINE INTELLIGENCE (1950–1952)	• He came up with the Turing Test in his paper titled "Computing Machinery and Intelligence". • The test sought to answer the question: "can machines ever think like humans?" • Three Laws of Robotics were formulated by Isaac Asimov.
JOHN MCARTHY—THE FATHER OF AI (1955–1956)	• Created the term "artificial intelligence". • Organizer of the first ever AI conference, the Dartmouth Conference of 1956. • The aim was to find out if a machine could be made that could think dynamically, resolve issues, and develop itself like a human. • Human-level AI and realistic thinking were two of his significant commitments.
MARGARET MASTERMAN (1960)	• Advancement in the field of AI developed quickly. • The making of new programming/computing languages, robotics and machines, research studies, and films depicting deceptively intelligent creatures expanded in notoriety. • Focused on the importance of artificial intelligence in the second part of the twentieth century. • Semantic nets were drafted for machine translation by Margaret Masterman and her co-workers at the University of Cambridge.
UNIMATE ROBOT (1961)	• The main modern robot, Unimate, began dealing with a sequential construction system and worked on General Motors' automobile assembly line. • Weight of its arm was 4,000 pounds. • Symbolic automatic integrator (SAINT) was developed by James Slagle
MACHINE LEARNING (1962)	• Arthur Samuel invented the term "machine learning" (ML). • He showed how ML can be utilized to play checkers.
FIRST CHATBOT—ELIZA (1963–1965)	• Developed by Joseph Weizenbaum. • An early NLP computer program • Created at MIT AI lab. • Established communication between humans and machines.
NEGATIVE FAME OF ML, SHAKEY ROBOT (1966–1969)	• The ALPAC report prompted the retraction of all administration-financed MT projects. • At Edinburgh, the first workshop on ML was held by Donald Michie. • The first international joint seminar on AI (IJCAI) was scheduled at Stanford. • A WABOT project was started at Waseda University, Japan. • Discussion between McCarthy and Hayes was held on the "frame problem". • The robot SHAKEY was designed by Charles Rosen. • Stanley Kubrick's sci-fi movie, *2001: A Space Odyssey*,, was released.
FIRST AI WINTERS (1970–1980)	• Due to the shortage of funding in the field of AI, most organizations stopped investing in AI-based surveys and research. • Overoptimistic guarantees by researchers, unnaturally exclusive requirements from end clients, and broad advancement in the media.
END OF AI WINTER (1981)	• XCON was developed by Digital Equipment Corporation.
FIFTH-GENERATION COMPUTER (1982)	• A project (FGCS) started by Japan's Ministry of International Trade and Industry aimed to perform logic programming and parallel computing. • Used for further development of AI.

TABLE 2.1   (Continued)

Brief History of AI [9]

Year	Description
A FUNDED INITIATIVE (1983)	• Strategic computing was set up by the US government to provide funded (DARPA) research in advanced computing and AI.
FURTHER DEVELOPMENTS (1984–1985)	• A team at Bundeswehr University, Munich developed the first ever robotic car which drove at a speed > 55mph. • At AAAI National Conference, an autonomous drawing program titled AARON was created by Harold Cohen.
AI PROGRAMMING LANGUAGE RAN (1986)	• Special computers built by companies like Symbolics and Lisp Machines Inc. to use AI programming language Lisp.
SECOND AI WINTER (1986–1993)	• As the publicity regarding AI expanded, analysts expected the field to face another winter. • High assumptions could not be satisfied. • Due to funding issues the second winter began.
ADVANCES IN AI (1994–1996)	• Significant advances in every aspect of AI like ML, NLP, games, virtual reality, chatbots, etc.
DEEP BLUE BEATS CHESS LEGEND (1997)	• Garry Kasparov, a former world chess champion, was defeated by IBM's Deep Blue computer. • An RNN structural design (long short-term memory) was developed by Jürgen Schmid Huber and Sepp Hoch Reiter.
KISMET (1998)	• Dr Cynthia Breazeal designed the first robot that can simulate human emotions. • KISMET was the first ever humanoid robot. It could walk upstairs.
AIBO (1999)	• The first robot dog was developed by Sony. • AIBO, the first AI pet. • Use of the World Wide Web was increasing rapidly.
EARLY (2000)	• "Year 2000 problem", also called the millennium bug, affected many computer systems. • ASIMO, an AI humanoid, developed by Honda.
FUTURISTIC MOVIE (2001)	• Release of the sci-fi movie *AI Artificial Intelligence.* • Smart toys available commercially.
ROOMBA: THE VACUUM CLEANER (2002)	• ROOMBA could detect dirty objects.
AI ROBOTS (2004–2005)	• A challenge was introduced by DARPA to make independent vehicles. • NASA launched two AI robots on Mars to explore the Martian surface without any human involvement. • The DARPA Grand Challenge was passed by five vehicles.
DATASETS CAME INTO THE PICTURE (2006–2007)	• "Machine reading" was defined by some researchers as "autonomous understanding of the text". • A dataset named ImageNet was built by Fei-Fei Li.
TECHNOLOGY ADVANCING (2008–2009)	• Google launched its self-driving car. • Big boom in intelligent software.
A REVOLUTIONARY ERA (2010)	• A body movement tracker was launched by Microsoft as a gaming device named XBOX 360. • Technology became cheaper and advanced at the same time. • More and more software and machines created. • Language recognition, speech translation, text generation, IoT, deep learning were developing at an enormous rate.
START OF NEW DECADE (2011)	• IBM created a human language-answering computer named WATSON. • UBER reached outside the US. • Different virtual assistants launched by different companies: Google launched Google Assistant, Apple launched Siri, Microsoft launched Cortana.

(Continued)

TABLE 2.1   (Continued)

Brief History of AI [9]

Year	Description
AI CHANGING THE GAME (2012–2013)	• A deep learning project by Google Brain was founded by Andrew Ng. • Carnegie Mellon University developed a NEIL (Never Ending Image Learner). • Cloud computing became a promising area.
ALEXA LAUNCHED (2014)	• Amazon launched virtual assistant or smart speaker named Alexa.
AI APPLICATIONS ON TOUR (2015–2017)	• 3,000 people, including Elon Musk, Stephen Hawking, and Steve Wozniak, signed a petition opposing the implementation and deployment of autonomous weapons. • World champion Go player Lee Sedol was defeated by Google DeepMind's AlphaGo. • A conference was held in California to discuss AI ethics. • A Hong-Kong-based company developed Sophia, a humanoid robot.
AI IN CHARGE (2018)	• Self-driving taxis used in Arizona and Phoenix. • Google-Duplex launched to book appointments via mobile phones.
EMERGING TECHNOLOGIES (2019)	• Smart homes and home automation gaining popularity. • Biometrics a hot research topic. • A talking hearing aid developed for people with hearing aids.
WAR BETWEEN COVID AND AI (2020)	• AI used in medicine. • 5G came as a big picture. • Covid-19 increased the use of AI in every field: education, healthcare, sports, agriculture, etc.
EARLY (2021)	• Samsung robotic arm • Predicting the next pandemic and many more to come.

2.2.3 Applications of AI

There are many applications of AI today, from voice assistants like Siri, Cortana, and Alexa to chatbots, from mobile phones to diagnosing diseases, from AI-based drones to AI-based automatic warehousing supervision systems. Fields in which AI is used include E-commerce, education, healthcare, automobiles, agriculture, gaming, sports, manufacturing, security, and robotics [13]. The field of AI is currently at the peak of its hype cycle. It is breathtaking, promising, and a bit frightening all at the same time. Many researchers believe that AI already recognizes what we want to purchase, it can produce a web series on Netflix, it can cure. As the adoption of AI becomes more widespread, AI-powered businesses may find it more difficult to maintain a competitive advantage over their sector counterparts. Most adopters expect AI to be instegrated into more and more generally available apps soon, indicating a levelling of the playing field. The applications of AI do not merely ensure good business performances; they also refine human gratification (Figure 2.1).

The primary phase of AI in education has consisted in design and provisional exploration. According to a recent survey, by 2025 the adoption of AI in education is estimated to cost $6 billion worldwide [14]. While the rapid progress of AI and its use in education are awe-inspiring, we should be mindful that systematic, reliable, and data-driven research and reports on the newest trends are vastly underrepresented.

For more than 30 years, academic research has focused on the application of AIED. To promote formal education as well as lifelong learning, this discipline examines learning wherever it occurs, whether in conventional classrooms or workshops. While the majority

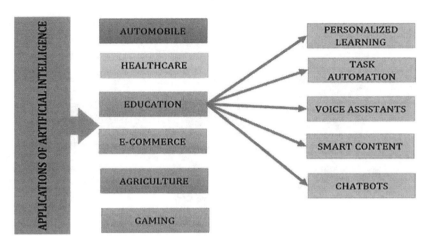

FIGURE 2.1
Applications of AI [13].

of AI research is conducted in STEM subjects, the growth in AIED applications necessitates multidisciplinary methods. With the arrival of the intelligence era, deep integration of artificial intelligence and traditional education has become a major trend in education. Companies that are following the "AI + Education" model are using the benefits of AI technology to focus on research and development in intelligent educational goods and services. With AI technology, it is feasible to teach pupils according to their ability, shifting the educational aim from high scores to quality improvement. With AI-based education still in its early stages, many people worry whether educators will be replaced by robots, whether the development will have beneficial or adverse consequences, and what can be done to enhance present teaching methods.

In our schools and colleges, a variety of AIED-based apps are already in use. AIED and educational data mining (EDM) technologies are being used to monitor student behavior, for example, gathering information on student engagement and project completion to identify (and assist) students who are at risk of dropping out. Other AI analysts are investigating innovative user interfaces such as natural language processing (NLP), voice and motion recognition, eye tracking, and other biological indicators [15]. AIED software applications that have been designed to explicitly engage learners include teaching assistants for each student, cognitive support for collaborative tutoring, and adaptive augmented worlds.

The use of AI in student acquisition and assistance via evaluation, training, and employment is composed of five key applications. We see robust advancement in a portion of these technologies with vital participants developing inside and across business sectors as academic institutions embrace AI-driven administrations and report positive effects from implementation. Computing machinery and intelligence is also deployed in educational games and applications to assist students in learning the material. For example, recent developments in technology allow video games such as *Minecraft* to be used as educational tools. These tools, in meeting the desire for more customized and tailored education content, allow teachers, parents, and caregivers to create an immersive environment where students can learn at their own pace. Personalized tools can assist teachers to spend less time on administration and more time on educating [16]. Advanced AI analytics are

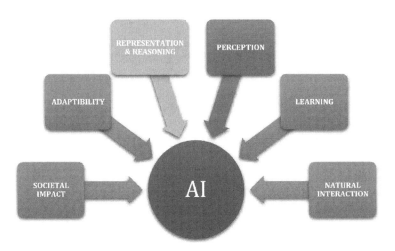

FIGURE 2.2
Ideas/Iimportance of AI.

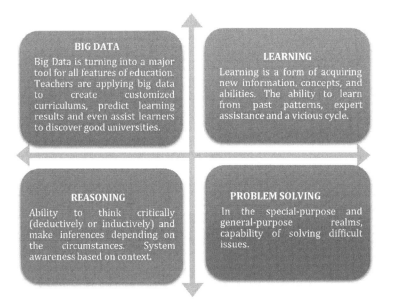

FIGURE 2.3
Key characteristics of an AI system.

changing the global educational environment, addressing some of the world's oldest problems, such as differentiated learning (Figures 2.2 and 2.3).

The six "big ideas" of AI in education are:

- *Natural interaction:* To engage properly with people, AI agents require a wide
  range of information. Agents should be able to communicate in human languages,
  identify facial feelings or expressions, and infer intents from observed behaviour
  using cultural and social standards. These are all challenging issues. Examine
  how AI detects human emotions by interacting with chatbots. Build a chatbot

and showcase it by controlling robots using voice control. These changes must be incorporated into Al's algorithms, and judgements must be made on how to adapt to the emerging technologies [16].

- *Learning:* Data may be used to teach computers. Machine learning is a statistical inference technique for detecting patterns in data. Learning algorithms that generate new representations have made considerable advances in several areas of Al in recent years. Massive volumes of data are necessary for the method to succeed. Understand the art of data in Al and try your hand at machine learning. Use a computational model to operate a robotics system to take it to the next stage [16].

- *Perception:* Sensors are used by computers to perceive the surroundings. The act of deriving meaning from sensory data is known as perception. One of Al's greatest contributions to the world utilizing sensors is enabling machines to "see" and "hear" well enough for practical usage[16].

- *Representation and reasoning:* Agents keep track of world depictions and utilize them to make decisions. One of the core difficulties of natural and artificial intelligence is depiction. Data structures are used by computers to build representations, and these representations are used by logical algorithms to infer new knowledge from what is previously known. While Al-based agents are capable of reasoning through complicated issues, they may not "think" in the same way that humans do [16].

- *Adaptability:* The capacity to learn and adapt while compiling data and making choices is the characteristic that distinguishes Al systems. As technology advances, effective artificial intelligence must adapt. Changes in financial status, traffic conditions, environmental concerns, or military situations may all be factors [16].

- *Societal impact:* The impact of AI on society can be both positive and negative, depending on different factors. How we work, drive, interact, and care for one another is all changing as a result of technological advancements. However, we must be aware of the risks that may arise. Preconceptions in the training and testing of an Al system might result in certain individuals receiving poorer service than others [16].

Here are Applications of AI in education that have gained traction in the last two years are:

- *Personalized learning:* AI-based personalized learning may help every individual to pick the quality education for their career trajectory, achieve their full potential, and acquire the skills they need to succeed in the digital workplace [17].

- *Artificial intelligence (AI) integrated with augmented reality (AR) and virtual reality (VR):* In education, augmented reality (AR) and virtual reality (VR) provide an artificial computer environment, something students can readily experience and engage with. Material can be transferred to students quickly and effectively. An engaging learning experience may open a whole new arena of fascinating learning possibilities [17].

- *Intelligent tutoring systems (ITS):* A computerized learning environment that enables students to master information and skills via the use of adaptive algorithms that adjust to individual students with a high level of precision and execute complicated learning concepts [17].

- *Adaptive learning:* A technique that adapts educational information to a student's active learning and speed using computerized artificial intelligence algorithms. Algorithms identify patterns in a student's response to information and reply on a real-time basis with prompts, modifications, and interventions adapted to the student's specific requirements and skills [17].

2.2.4 Visions and Challenges of AIED

Huge strides have been made in the field of education in the last two years, with new possibilities opening up for creating effective teaching and learning strategies and improved technology-enhanced learning apps or spaces. However, most scholars and experts in the domains of machines and education still find it difficult to put relevant activities or systems in place. Emerging technologies (such as image recognition, voice recognition, knowledge-based systems, and natural language processing) can be implemented on various operating systems, platforms or devices (including smartphones, wearable technology, and robots/automatons) to meet academic or training design objectives (such as issue-enhanced learning, key concept-based tutoring, and investigation-grounded learning) for different courses like engineering, social science, art, design, medicine, mathematics, algorithms, etc. [18]. Some learners and speakers have a negative attitude to these frameworks, regarding them as not appropriately adapted to their particular specialist requirements. This has led to further development of learning frameworks. Artificial intelligence integrated with chatbots can help students and educators in more intelligent interaction. The ideal application of AI in education is to assist instructors to deliver more productive classroom instruction by augmenting teacher capability. Marking and dealing with student queries is a time-consuming task that could be optimized by educators using AI. AI is also assisting teachers in pointing out curriculum areas that need updating. Not only this, but it is also empowering various education institutions to devise customized education for their students. Moreover, it helps to monitor and evaluate students' learning speed and needs. Education is now facing the challenges of using AI to enhance educational practice both in the classroom and at the core level, and engaging learners for emerging technical skills in more computerized societies and economies. The example of China shows four technical challenges remain in the current rapid improvement of education digitalization and intellectual ability: "normal" students' individual characteristics are not completely considered in education, and the integrity of normal education must be enhanced; teacher education needs to be improved; there is still a lack of complete knowledge of students' customized features; and personalized instruction requires more effort. Our analysis suggests that the three challenging technological obstacles are technique, instructors and students, and social ethics [19] (Table 2.2).

2.2.5 EdTech Start-Ups

With huge investments in nations like China, the United States, and India, the education tech industry, often known as EdTech, is booming. From digital platforms to robotics and smart gadgets, it creates a wide spectrum of innovative tools for educational institutions and stakeholders. Artificial intelligence (AI) is reshaping the world, including education. $3.67 billion was invested in AI education technology start-ups in 2019, up from $2.89 billion in 2018 [21]. Technological giants like Google, Microsoft, Apple, IBM, and Baidu are among the world's most powerful technological corporations. These tech giants are keen to investigate the potential of AI for educational purposes, armed with an immense

TABLE 2.2

Challenges and Benefits of Adoption of AI [20]

Challenges	Benefits
• Accessibility to technology and the internet is the biggest concern many countries across the globe are still confronting. • Data security is among the top threats to AI • Diminishing the creativity of students, as everything can be done using the software. • Distracts students while attending lectures. • The use of AI leads to an increase in mental health problems among both teachers and students. • Absence of human touch or teachers. • Data management • Planning instructors for AI-fuelled schooling and training • Affordability • Lack of positive approach for AI • Increases unemployment, as classes are taken by robots. • Increase in screen time for students	• Personalized learning, adaptive learning, intelligent tutoring systems. • Adoption of EdTech to rebuild education with AI and analytics. • During the pandemic, AI was the only way of learning, as schools, universities, and libraries all closed. • Hyper-personalization • Virtual and augmented reality • Improved content quality by creating smart content and improved decision making. • A game-changer for disabled students. • Virtual assistants helping with language barriers and a 24/7 teaching assistant • Reduction in human error. • Ease in handling repetitive tasks and free time for more creative work and hobbies. • Essay grading software • Detecting educational abnormality

amount of data and top expertise in machine learning (ML) and data science [22]. Google AI education has a collection of AI applications and experiments, as well as guidelines, training resources, and courses for beginners and experts alike. For individuals interested in pursuing AI as a profession or education, Microsoft AI School also provides similar pathways and learning routes. Apple Education and IBM Watson Education are two further instances of tech titans exploring AI-based student engagement and boosting active learning. Tech giants in the digital realm are typically dominated by intelligent personal assistants such as Apple's Siri and Amazon's Alexa in the NLP bot paradigm [23]. Several recently established AI firms are also included in the list of prominent technological companies. SenseTime, iFlytek, and AIBrain are a few worth noting. Such AI firms are also expanding their operations into the field of education, either by experimenting with novel K-12 teaching techniques or by establishing AI colleges to train fresh talent [24].

In contrast to the confusion and damage caused to various sectors—including education—by the Covid-19 global pandemic, the accelerated penetration and rapid adoption brought on by lockdown has been positively aiding the growth of the EdTech industry, according to statistics. Studies reveal that the EdTech category experienced a spectacular 26% rise in user interactions during Covid-19, with prominent EdTech firms claiming a 100% gain in paying users month over month and a 50% increase in traffic [25] (Table 2.3).

Figure 2.4 shows emerging technological applications in education and research which by 2025 will have reached a tipping point, with AR/VR and AI being more integrated into fundamental quality education and educational experiences [26].

2.2.6 Education during Covid-19

The Covid-19 pandemic has caused a tectonic shift in teaching and workplace norms. Universities had to quickly adjust the available technology to an online learning model and increase their technological capacity, communications, and digital security. This topic

TABLE 2.3

India's top EdTech companies

Edtech Company	Focus Area	Specialty	Features
BYJU'S	Personalized learning app, Programmes for K-12, and aspirants to competitive exams like JEE, IAS, etc.	Exciting and highly engaging learning modules include unique material, watch-and-learn lectures, animations, and interactive scenarios that help students comprehend concepts in a fun and exciting way.	• For both children and adults • Accessible for learners across grades. • Frequent scholarship tests
TOPPR	Personalized online learning app for CBSE, ICSE, and State Board students.	Mixing high-quality content with cutting-edge technology and design	• Adaptive practice, Live Q&A with expert tutors
VEDANTU	For classes 6–12. Mathematics, science, social studies, Hindi, German, French, English, etc.	Personalized training delivered live to the convenience of your own home.	• Free live lectures and classes (two-way teacher–student connection). • Live Q&A, lessons with live assessments. • Free learning materials, NCERT explanations with solutions. • Past exam papers, key questions, learning monitoring tools, and performance results.
UNACADEMY	On SSC, UPSE, STATE PSE, banking, railways, defence IIT & AIMS exams	Broadcasts live classroom videos from exceptional educators, which are also available offline. While every student has unrestricted access to faculty members to ask questions, time may be an issue.	Suitable for competitive exams. Live classes, live mock tests for practice.

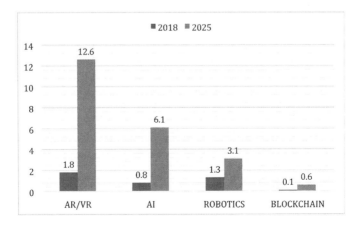

FIGURE 2.4

Advanced EdTech expenditure 2018–2025, in US$ billions [26].

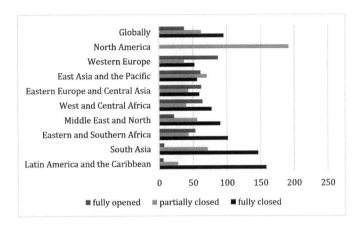

FIGURE 2.5
No. of students (in millions) impacted across different countries during Covid-19 [22].

will continue to be a priority for the foreseeable future, as the trend is to heavily focus on new technologies, which will translate into a need for higher levels of service and cyber-security [27].

Digital skills are crucial to help organizations adapt to short-term crises and promote long-term transformation and skill development. Retraining the workforce is becoming more of a priority. One way that businesses are reacting to the problem is by adopting and adapting AI. While Covid-19 caused a general slowdown in the economy, the current crisis has boosted the use of online platforms, particularly in consumer-facing industries. Although traditional education was damaged, the Covid-19 lockdowns stimulated the growth and investment of the Indian EdTech industry. According to a UNESCO survey, by May 2020, more than 1.2 billion children and adolescents had stopped taking face-to-face classes due to the Covid-19 crisis [28]. The data analytics and AI frameworks created by worldwide innovation firms and EdTech organizations have become trial motors of algorithmic learning and educational systems have become their research facilities for new computerized types of instructing and learning in the post-pandemic future (Figure 2.5). Several big companies jumped on the post-Covid possibilities by diversifying into new product categories in biotic and abiotic ways. Across the world, the integration of distance learning methodologies through a range of categories and platforms (with or without the utilization of emerging technologies) has been accompanied by greater attention to students' wellbeing, and the assistance and mobilization of education staff and organizations [29].

2.3 Literature Survey

During the last 60 years, numerous surveys and research on the subject of AIED have been conducted and significant developments have occurred since then. AI development, logical thinking, and qualitative temporal conceptual characteristics have all been studied by researchers. Mobile technology, cloud computing, big-data analytics, and substantial advances in artificial intelligence have all had a major influence on education [30].

In recent years, more advanced AI-based training techniques have evolved, and they are gaining momentum owing to their capacity to provide learning material and acclimatize to learners' specific requirements. Although these optimization techniques are effective educational tools that meet students' demands, there remain limited systems in place that solve the issues and challenges that many students experience. Premised on this viewpoint, a comprehensive survey of prestigious publications on AI-based adaptive learning techniques was conducted for this chapter. We reviewed more than 120 papers published between 2015 and 2021. In current literature, we found three key AIED paths: learning with AI, learning from AI, and learning about AI. Previous studies raised several worries about online learners encountering obstacles once they are in the virtual context, which necessitates the use of specialized abilities to engage with and acquire e-learning resources.

The limitations of adopting new technology in the classroom as a strategic change issue are explored in a wide body of scholarly research. In addition, adverse responses to technology in higher education have a long history in the educational literature [31].

Many research articles exist in this field, although most of it was conducted by computer scientists (using data from genuine educational institutions) and has yet to be completely applied within universities. With the proliferation of online publications and open-source services, doing a comprehensive search, even with well-defined constraints, is almost impossible. Our research was therefore deliberately refocused on research articles found in the most extensively used web-based databases like Google Scholar and Web of Science [32] (Figure 2.6 and Table 2.4).

2.4 Findings and Discussion

This section summarizes the findings based on our review of the scientific research. Figure 2.7 presents a methodical flowchart highlighting the key elements of this review, based on prior work in this field.

Figure 2.8 shows the pattern of research publications between 2017 and 2020 that used the abovementioned AI-based learning subjects [34, 35].

2.5 Conclusion

The adoption of artificial intelligence is essential for promoting transformational innovations that use automation, support, and personalization to improve human experiences. This qualitative research used a literature survey to analyze the influence of AI on education. The chapter analyzes and presents a variety of viewpoints in the study of AI in education. The conclusion is that policies to regulate and supervise the use of innovation are needed to achieve an active and productive result. For example, ethical problems raised by intelligent systems should be handled by regulation and standards that identify who is accountable for the data collected by the system. As AI develops alternative predictive paradigms, the difficulty for education research in the 2020s will be to critically interrogate assumptions and unresolved conflicts in order to avoid

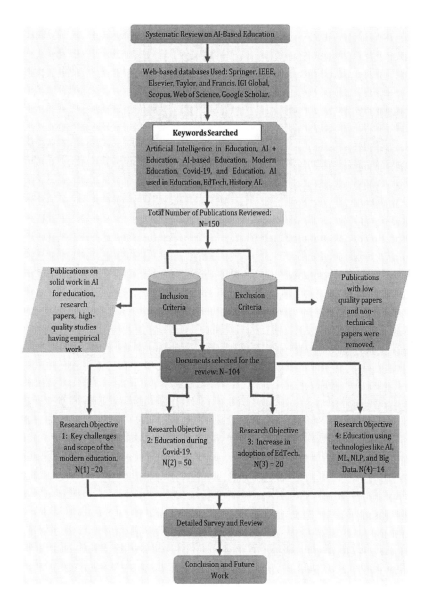

FIGURE 2.6
PRISMA for the systematic literature review.

complete ineffective closure. Altogether, AI has had a substantial influence on education, notably in the expansion of management, training, and teaching in the education sector or within specific educational institutes. Current worldwide innovative ideas cannot be attributed simply to the third industrial revolution; rather, they mark the start of a fourth industrial revolution in which a variety of technologies are blurring the lines between the digital, physical, and physiological domains. We have visualized and conceptualized the domain of AI and education and featured how personalization in AI is a key idea for education reform.

TABLE 2.4

Literature Survey

Reference no. / Author	Proposed Work	Relevant Findings	Future Work
[6] Camelia Şerban et al.; 2020	• The motivation behind this investigation is to integrate Alexa into online learning.	• The working paper proposes a software application(Alexa for Uni) using the fusion of Alexa and some services (web services, Microsoft services). • The smart bot(Alexa for Uni) incorporates other significant administrations for the Covid-19 pandemic.	• A more intuitive user-friendly interface for gadgets with panels is at the top of the list. • To use Alexa for Uni for business from Amazon Web Services to use cloud services; for exam preparation, using MS Word and OneDrive.
[7] A. M. Cox et al.; 2021	• To investigate the influence of AI and machines on higher education. • A literature survey was conducted to develop eight stories that capture the scope of possible utilization of AI and robotics in knowledge, organization, and exploration.	• Fiction is used in discussion among students, to investigate the effect of AI on education. • The paper has clarified the reasoning and interaction of composing fiction.	• Some more futuristic fiction should be developed to fill the gaps of development.
[8] Xieling Chen et al.; 2020	• The working paper aims to investigate the impact of AI on education. • The author evaluated 45 journal articles and illustrated the definitions of AI in education in EDM, machine-based learning, and learning analytics.	• AI is showing the impactful nature of education. • Among the 45 AIED articles, 41 relate to machine learning integrated with education, while just two have a place with DLED. • Deep Learning and NLP are two of the most widely used methods for AIED studies.	• Potential of utilizing AI in classrooms. • Future studies envisage looking at AIED research in some countries to acknowledge crucial organizations and researchers from a national perspective. • Within the huge scope of AIED research, the utility of text mining and NLP approaches.
[9] Aras Bozkurt et al.; 2021	• Aim to review the literature on AIED between 1970 and 2020. • To recognize research patterns and examples of AI in educational publications.	• They reviewed 276 publications with 144 conference research articles and 132 normal articles. • The study shows the geographical distribution of the publications available on AI in education across the globe. • It also illustrates the relationships among technologies like AI, ML, and DL in relation to education. • Demonstrates trending in AI research in education. • They examine the different subject areas of publications on AIED.	• Rise in cross-cutting studies, since the distribution of AI in instructive settings cannot be led without the assistance of specialists in education and sociology because of the need to plan AIED applications with combined information on hypothesis and practice in schooling. • The ideal would be to accept that AI would be utilized freely by disregarding elements in the field of education.

[10] Mehrnaz Fahimirad et al.; 2018	• The paper reviews the various available teaching and learning applications of AI in education. • This reasonable survey paper groups articles on coordinating AI in education by ideas and topics. • It depicts a snapshot of the future of AI in education.	• Showcases the available AI tools used by students. • Discusses challenges and opportunities of AI in education. • Reviews the integration of the human mind with AI.	• Researchers and scholars need to consider this aim to foster responsible people and educated thinkers.
[11] Fati Tahiru et al.; 2021	• This article examines the prospects, benefits, barriers, and challenges of AI in education. • An in-depth study on the current literature review was done to find a relevant future research roadmap. • They aim to depict the geographical area in which the study of AIED is concentrated.	• AI is impacting education positively. • Some opportunities are developed for students by collaborating in different learning techniques. • ITS is a highly recommended AI tool. • Very few pieces of literature were available during 2010–2015 but a rapid increase in the literature was seen between 2016 and 2019. • To explore the problems of AI in education, they embraced the TOE framework as a standard.	• There is a need to evaluate the key gains and potential of AI in face-to-face classrooms. • Further research should tackle the importance of the integration of big data and AI to open up new research topics. • Work should be done on the ethical issues of AI in education.
[12] Chong Guan et al.; 2020	• They study research articles from 2000 to 2019 to identify historical and recent research trends. • They reviewed 400 research articles to examine the application of DL and AI in education.	• The outcomes recommend a decrease in regular tech-empowered informative plan investigation and the prospering of understudy outlining patterns and learning analysis. • This study sheds light on the opportunities and difficulties behind AI and DL for academic transformation and starting a discourse.	• Future studies should include the concept of EdTech in education branding.
[13] Stephen J.H. Yang et al.; 2021	• This paper explores the utilization of AI to assess new plan techniques and instruments that can be utilized to boost advanced AI research, instruction, strategy, and practice to benefit mankind.	• They investigate how AI can hinder the human condition and support comprehensive exchange among innovation and human analysts to improve comprehension of HAI from different points of view.	• Analysis of data and acquiring a good job will be more important in the future employment market. • Future job seekers will need to have a lot of specialized abilities and preliminary information, particularly information and ideas identified with AI.

(Continued)

TABLE 2.4 (Continued)

Literature Survey

Reference no. / Author	Proposed Work	Relevant Findings	Future Work
[14] Tony Bates et al.; 2020	• What are the reasons for research ignorance in the field of HE in AI? • What are the unmet needs of education in terms of learning and teaching? • How can AI innovation help diminish existing and future disparities?	• It will be increasingly important for advanced education institutes to become agile learning associations prepared to rapidly embrace new practices and elements. • In recent years, the greatest development in AI has been expansion of new interfaces.	• Regardless of whether innovation supplants teachers through automation, or whether innovation ought to be utilized to enable not just instructors yet additionally students. • Most importantly, who should control AI in training: instructors, understudies, PC researchers, or huge companies?
[15] Fan Ouyang et al.; 2021	• In this paper three paradigmatic shifts are categorized. • To sum up the significant standards with the portrayals of important hypothetical establishments, reasonable examination, and viable executions. • In the present development-driven age, it provides a reference framework for future AIED practices, investigation, and progress that can support student-focused learning and big data.	• In Paradigm 1, AI frameworks anticipate psychological educational programmes, while learners rely on AI administration to keep track of their progress. • In Paradigm 2, the AI structure and the learners collect shared correspondences, resulting in more student-centred learning. • To generate adaptable, personalized learning, Paradigm 3 emphasizes the synergetic association, mix, and coordinated effort between the AI framework and human knowledge.	• We contend that the future advancement of the AIED field should prompt iterative improvement of student-focused, information-driven, customized learning in the current information age.
[16] Shalini Garg et al.; 2020	• The focus is to study the impression of AI on students with disabilities. • They demonstrated the problems associated with children with special needs.	• The study recommended a comprehensive teaching method that incorporates each youngster without arrangement. • It additionally proposes creating teaching methods that advance imagination and will help establish a protected environment for children where they are allowed to trade thoughts, discuss with one another, and recognize each person's differences.	• Further research should focus on AI-based gadgets which are used to sense disability. • The educator needs to foster instructional methods that cause children to have a sense of security and be allowed to trade their thoughts and discussions with one another. The teaching methods are planned in a manner that advances innovativeness.

Reference			
[17] Marcel Pikhart et al.; 2020	• To bring out new ideas for the use of language and vocabulary learning apps. • This paper tries to examine the reason behind the neglect of learning apps. • How AI and ML can be used to enhance the usage of these apps. Research is focused on 10 foreign language learning mobile apps	• Most of the apps are not using AI, ML, and DL, but are showing their predefined algorithm. • Highlights the importance of the use of AL, ML, and DL in mobile learning apps.	• Advanced education organizations ought to be prepared to assist with the turn of events of AI-empowered gadgets and applications with the goal that they can be utilized in their educational process.
[20] LIJIA CHEN1 et al.; 2020	• To identify the impact of AI on education in different forms. • They provide some techniques for the framework of AI in education. • They elaborate on the role of AI in education, learning, administration, and instruction.	• A qualitative research survey. • Assisted by learning analytics, AI, ML, and data mining, frameworks will give top-notch content to instructors and students, to help both educating and learning and make the entire cycle quantifiable	• As the research and development in the field of AI is rapidly increasing there is a huge scope for study in the context of education and learning.
[19] Eyman Alyahyan et al.; 2020	• This study aims to fill the gap identified by providing a comprehensive prescription, making data mining methods more accessible, and enabling the full potential of their application to education. • This analysis presents an unmistakable arrangement of rules to follow for utilizing EDM to forecast progress.	• It is possible to create forecast models to increase student success using EDM techniques. • For non-technical individuals, using data mining methods might be intimidating and difficult.	• It is a major consequence of this survey that teachers and non-capable users are urged to apply EDM strategies for college students.
[21] Kashif Ahmad et al.; 2020	• They reviewed various AI applications in education using different approaches. • They provided a detailed literature survey from the year 2014–2019. • They also provide a bibliometric analysis of techniques used in AIED.	• They address research questions such as: • How will technology pop-up learning in structured new ways affect life expectancy? • How far can we go towards adaptive learning? • How might new information and techniques for examination uncover pathways to progress?	• Analysts are deprived of fostering wise AI strategies that are prepared to manage information in classified and cautious ways. • Computer-based intelligence analysts need to search for approaches to refine their calculations and examinations with regard to dissecting information and identifying designs.

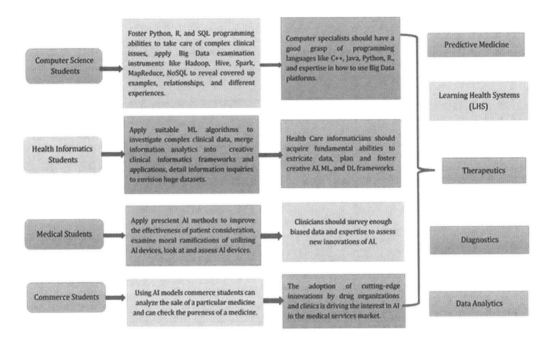

FIGURE 2.7
Proposed framework on how AI can be used by different educational subjects in the healthcare sector [33].

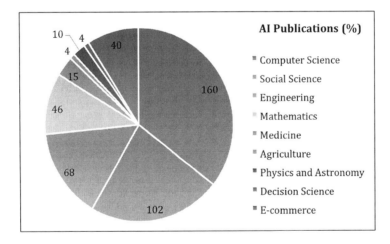

FIGURE 2.8
Classification of the areas of the publications on AI in education [34].

2.6 Future Work

In the past few years, the perception of artificial intelligence (AI) has shifted from classic rule-based or analytical learning techniques to deep learning methods, owing to the rapid advancement and substantial success of different deep learning approaches in AI. This shift in AI has resulted in substantial breakthroughs in both the academic and industrial sectors. Further investigation is needed in this multidisciplinary field concerning the use of technology in education in real-world settings, with an emphasis on methodology or administrative procedures rather than the technology itself. To comprehend the potential influence of AI evolution on upcoming education and learning, it is important to re-evaluate the possibilities, and study the objectives and responsibilities of inclusive education, as innovative AI empowers vital roles in education that are not confined to intuitive mentors, efficient learning associates, and policymaking advisers. Further analysis in the specific domain of technology for education must be emphasized. Administrators must modify educational policies at different levels (beginner, intermediate, and advanced) to inculcate the fundamental knowledge needed to design or construct AI applications in students and assure AI's long-term viability. An unbiased and objective research study is required to establish the effects of sophisticated AI-based technologies on teaching and learning. The analysis in this research does not seek to be exhaustive but rather to provide a systematic outline of AI in education. And it may provide a foundation for future research integration. Furthermore, future research may go back in time to determine whether there were any modifications around the time AI 2.0 began to make inroads into education.

Note

1 https://www.ibm.com/downloads/cas/3RL3VXGA.

References

1. Yang, S., & Bai, H. (2020). The integration design of artificial intelligence and normal students' education. *Journal of Physics: Conference Series*, *1453*(1), 012090. IOP Publishing.
2. Cope, B., Kalantzis, M., & Searsmith, D. (2020). Artificial intelligence for education: Knowledge and its assessment in AI-enabled learning ecologies. *Educational Philosophy and Theory*, *53*(12), 1229–1245.
3. Dewi, R. R. V. K., Muslimat, A., Yuangga, K. D., Sunarsi, D., Khoiri, A., Suryadi, S., & Iswadi, U. (2021). E-learning as education media innovation in the industrial revolution and education 4.0 Era. *Journal of Contemporary Issues in Business and Government*, *27*(1), 28682880.
4. Popenici, S. A., & Kerr, S. (2017). Exploring the impact of artificial intelligence on teaching and learning in higher education. *Research and Practice in Technology Enhanced Learning*, *12*(1), 113.
5. History of Artificial Intelligence. (2022) https://en.wikipedia.org/wiki/History_of_artificial_intelligence
6. Șerban, C., & Todericiu, I. A. (2020). Alexa, what classes do I have today? The use of artificial intelligence via smart speakers in education. *Procedia Computer Science*, *176*, 2849–2857.
7. Cox, A. M. (2021). Exploring the impact of artificial intelligence and robots on higher education through literature-based design fictions. *International Journal of Educational Technology in Higher Education*, *18*(1), 1–19.
8. Chen, X., Xie, H., Zou, D., & Hwang, G. J. (2020). Application and theory gaps during the rise of artificial intelligence in education. *Computers and Education: Artificial Intelligence*, *1*, 100002.
9. Bozkurt, A., & Sharma, R. C. (2020). Education in normal, new normal, and next to normal: Observations from the past, insights from the present, and projections for the future. *Asian Journal of Distance Education*, *15*(2), 1–10.

10. Fahimirad, M., & Kotamjani, S. S. (2018). A review on application of artificial intelligence in teaching and learning in educational contexts. *International Journal of Learning and Development*, *8*(4), 106–118.
11. Tahiru, F. (2021). AI in education: A systematic literature review. *Journal of Cases on Information Technology (JCIT)*, *23*(1), 1–20.
12. Guan, C., Mou, J., & Jiang, Z. (2020). Artificial intelligence innovation in education: A twenty year data-driven historical analysis. *International Journal of Innovation Studies*, *4*(4), 134–147.
13. Yang, S. J., Ogata, H., Matsui, T., & Chen, N. S. (2021). Human-centered artificial intelligence in education: Seeing the invisible through the visible. *Computers and Education: Artificial Intelligence*, *2*, 100008.
14. Bates, T., Cobo, C., Mariño, O., & Wheeler, S. (2020). Can artificial intelligence transform higher education? *International Journal of Educational Technology in Higher Education*, *17*, 1–12.
15. Ouyang, F., & Jiao, P. (2021). Artificial intelligence in education: The three paradigms. *Computers and Education: Artificial Intelligence*, *2*, 100020.
16. Garg, S., & Sharma, S. (2020). Impact of artificial intelligence in special need education to promote inclusive pedagogy. *International Journal of Information and Education Technology*, *10*(7), 523–527.
17. Pikhart, M. (2020). Intelligent information processing for language education: The use of artificial intelligence in language learning apps. *Procedia Computer Science*, *176*, 1412–1419.
18. Zawacki-Richter, O., Marín, V. I., Bond, M., & Gouverneur, F. (2019). Systematic review of research on artificial intelligence applications in higher education—Where are the educators? *International Journal of Educational Technology in Higher Education*, *16*(1), 1–27.
19. Alyahyan, E., & Düştegör, D. (2020). Predicting academic success in higher education: Literature review and best practices. *International Journal of Educational Technology in Higher Education*, *17*(1), 1–21.
20. Chen, L., Chen, P., & Lin, Z. (2020). Artificial intelligence in education: A review. *IEEE Access*, *8*, 75264–75278.
21. Ahmad, K., Qadir, J., Al-Fuqaha, A., Iqbal, W., El-Hassan, A., Benhaddou, D., & Ayyash, M. (2020). Artificial intelligence in education: A panoramic review.
22. Hwang, G. J., Xie, H., Wah, B. W., & Gašević, D. (2020). Vision, challenges, roles and research issues of artificial intelligence in education. *Computers and Education: Artificial Intelligence*, *1*, 100001.
23. Bonfield, C. A., Salter, M., Longmuir, A., Benson, M., & Adachi, C. (2020). Transformation or evolution?: Education 4.0, teaching and learning in the digital age. *Higher Education Pedagogies*, *5*(1), 223–246.
24. Vincent-Lancrin, S., & Van der Vlies, R. (2020). *Trustworthy artificial intelligence (AI) in education: Promises and challenges*. Paris: OECD.
25. Zhang, K., & Aslan, A. B. (2021). AI technologies for education: Recent research & future directions. *Computers and Education: Artificial Intelligence*, *2*, 100025.
26. Williamson, B., & Eynon, R. (2020). Historical threads, missing links, and future directions in AI in education. *Learning, Media and Technology*, *45*(3), 223–235.
27. Verma, A., Kumar, Y., & Kohli, R. (2021). Study of AI techniques in quality educations: Challenges and recent progress. *SN Computer Science*, *2*(4), 1–7.
28. Chick, R. C., Clifton, G. T., Peace, K. M., Propper, B. W., Hale, D. F., Alseidi, A. A., & Vreeland, T. J. (2020). Using technology to maintain the education of residents during the COVID-19 pandemic. *Journal of Surgical Education*, *77*(4), 729–732.
29. Kushwah, A., Panda, S., Krishna, M. M., Alam, M. F., & Srivastava, A. (2021). Online learning in India: Growth, key drivers and challenges. *International Journal of Research in Engineering, Science and Management*, *4*(2), 129–132.
30. Khomova, O., Yanchytska, K., Shkatula, O., Burak, V., & Frolova, O. (2021). Trends in the development of tertiary education in the context of modern challenges. *Applied Linguistics Research Journal*, *5*(4), 126–133.

31. Rathore, N. P., & Dangi, M. (2021). Embedding Artificial Intelligence into Education: The New Normal. In *Applications of Artificial Intelligence in Business, Education and Healthcare,* ed. Allam Hamdan et al. (p. 255270). Springer, Cham.
32. Chaudhry, M., & Kazim, E. (2021). Artificial Intelligence in Education (AIED) a high-level academic and industry note 2021. *AI and Ethics,* 1–9. *Available at SSRN 3833583.*
33. Zhai, X., Chu, X., Chai, C. S., Jong, M. S. Y., Istenic, A., Spector, M., & Li, Y. (2021). A review of Artificial Intelligence (AI) in education from 2010 to 2020. *Complexity,* 1–18.
34. Toivonen, T., Jormanainen, I., & Tukiainen, M. (2019). Augmented intelligence in educational data mining. *Smart Learning Environments, 6*(1), 1–25.
35. Williamson, B. (2021). Education technology seizes a pandemic opening. *Current History, 120*(822), 15–20.

3

Multidisciplinary Applications of Machine Learning

Gnanasankaran Natarajan, Rakesh Gnanasekaran, and Manikumar Thangaraj
Thiagarajar College, Tamil Nadu, India

CONTENTS

DOI: 10.1201/9781003240310-3

3.1 Introduction

The term "machine learning" represents the automated disclosure of consequential data patterns. It has developed into a powerful device in relatively any function that needs knowledge to be abstracted from a very large dataset. Our real environment is surrounded by machine-learning technology. Search engines satisfy our needs by providing fantastic results, email messages can be filtered out using anti-spam software, and credit card transactions are secured using sophisticated software. Digital cameras can detect faces, and in smartphones voice commands can be easily recognized by intelligent personal assistant systems. Vehicles are fitted with accident anticipation systems. We can foresee a promising harvest in the precision agribusiness. Traffic light checks are mostly conducted using AI calculations [1].

Computers can learn automatically from previous data with the help of machine-learning technology. Mathematical representations and predictions can very easily and efficiently be formulated using powerful algorithms adopting archival data or information. Countless tasks are performed such as the realization of images, voice recognition, filtering emails, auto-tagging in social media websites, rule-based recommendation systems, and in scientific applications such as bioinformatics, medicine, smart farming, smart healthcare, weather forecasting and traffic signal monitoring. Common to all these applications is that in all these scenarios a human programmer cannot provide a definitive, detailed stipulation of how much work has to be executed. But machine learning can provide successful models and predictions to accurately carry out any tasks. As human beings, we gain experience and knowledge from the work we do and the concepts we learn. Machine-learning tools are equipped with modern programs through which they acquire the ability to learn and experience [2].

The primary aim of this chapter is to provide a detailed introduction to machine learning, a clear understanding of machine-learning concepts, an explanation of the situations where machine learning can be implemented effectively, and the classifications of machine learning.

Specific attention is also paid to artificial intelligence and big data, since we live in a digitized world where data available for learning is evolving continuously. Since the data is abundant, the computation time becomes more challenging. Therefore, we try to quantify both the volume of data and the quantity of computational time needed to learn a particular concept.

Finally, we discuss the various applications of machine learning in multidisciplinary fields of research such as precision agriculture and smart healthcare, as well as its technological implementations.

3.2 Machine Learning: A Prolific Concept to Make Machines Learn

In recent decades, humans have recognized the uses of everything through past learning and the knowledge they have acquired, and computers function according to the instructions provided. But we wonder whether a machine can also learn as humans do? Machine learning is a well-known subcategory of artificial intelligence that essentially deals with the expansion of algorithms enabling a computer to master understanding of previous data by itself [1].

Several mathematical models help develop predictions or decisions using classical data fragments, also called training data, without being explicitly programmed. Machine learning helps to shape these mathematical models to provide more accuracy and efficiency. Computer science and statistics integrated for machine learning can create predictive models. If we provide more information, the performance of the model also improves. If we feed the system with plenty of relevant data, it acquires the ability to learn more.

3.2.1 Machine Learning: Workflow

Machine learning builds predictive models by learning historical data, and every time it finds new data that is given in input, it predicts the result for it. Machine learning has changed our way of thinking about a particular problem and how to deal with it. Take the example of a complex issue, where we need to determine more than one output. Rather than creating a drawn-out code, we simply load the framework with information and certain calculations, and machine learning uses these to provide a clearer way of thinking according to the information given. Figure 3.1 depicts the workflow of a machine-learning algorithm:

3.2.2 Prominent Features of Machine Learning

- Machine learning trains the data to generate several arrangements for an inclined dataset.
- It can learn from former data and reform according to the amount of data given as input.
- It is a data-directed mechanism.
- Machine learning is analogous to data mining as it also handles a massive amount of data.
- It acts as a platform to bring out new technology like deep learning.

3.2.3 A Strong Understanding of Machine Learning

As humans, our ultimate objective is to compute modern systems so that they can train from the input fed to them. The transformation that remodels experience into expertise or wisdom is termed a learning process [1]. The system can gain knowledge and the more impressive and accurate the output produced from its model is, the more effort is invested in providing the training data. It usually takes the form of an additional computer program that can complete tasks accurately and precisely. In order to realize this concept, we need to familiarize ourselves with a few important facts such as: what kind of training data will

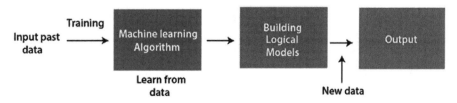

FIGURE 3.1
Workflow structure of a machine-learning model.

our system need? How can this learning process be automated? How will we be able to judge that the process has been completed successfully?

3.2.4 Machine Learning: A Tool Needed at the Right Time

When exactly do we need machine learning rather than precisely programming our systems to accomplish the work at hand? It is important to remember that ML is not a one-stop shop for every type of problem. There are certain situations where standard solutions can be determined without using ML techniques. Machine learning is not required for problems that already have predefined rules, principles, and genuine calculations.

Machine learning can be used in the following circumstances:

- *When rules cannot be coded:* Several manual tasks such as identifying whether an email is spam or not, predicting the yield of a crop, controlling or monitoring traffic, or identifying the risk of disease and health conditions in humans cannot be easily solved using a simple, common, rule-based solution. There is a large number of factors that could provide the correct answer. It is difficult for a human to precisely and accurately code the rules when outnumbered by rules and methods.

- *When scalability becomes tedious:* You might be able to manually track a few emails and decide whether they are spam or not. However, this task becomes impossible for millions of emails. You cannot easily predict the yield of numerous crop varieties over a wide area. You cannot control or monitor traffic over the entire district. ML solutions are effective for handling large-scale problems [1].

Other key factors indicating the need for machine learning are:

1. Unexpected growth of data production across domains.
2. Complex problems which are certainly not possible for a human to solve and find a proper solution using ML algorithms.
3. Standard decision-making in various fields of science, technology, finance, medicine and drug research.
4. By excavating useful information from the data provided, ML can identify some useful patterns [2].

3.3 Classifications of Machine Learning

Machine learning is divided into three broad categories: supervised learning, unsupervised learning, and reinforcement learning.

3.3.1 Supervised Learning

In supervised learning, we provide the machine-learning system with a sample labelled dataset to train it, and based on the data provided, the output will be predicted by the system.

Using perceptible information, the framework model is made to understand the datasets and decide on each piece of information. When the data have been satisfactorily established and refined, we review the plan by providing the framework with example information to check if it is gauging the exact yield appropriately.

Supervised learning aims to relate input data to output data. The mechanism behind supervised learning is proper oversight, and it is very similar to a student learning a subject with the help, supervision, and support of the teacher. One of the best examples of supervised learning is spam filtering.

Supervised learning is further divided into two classes of algorithms: classification and regression.

3.3.2 Unsupervised Learning

When a system tries to learn without any help, support, or further supervision it is termed unsupervised learning.

In this learning method, the system is provided with a dataset that may not be labelled, grouped, or typecast, and the algorithm acts on the data without any supervision. The main purpose of unsupervised learning is to remodify the input data into possible new features or a set of objects with coinciding patterns.

In unsupervised learning, we cannot get an expected result. Useful insights can be identified by the system from the massive quantity of data. This method can be further classified into two types of algorithms: clustering and association.

3.3.3 Reinforcement Learning

Reinforcement learning is quite similar to a feedback-based learning method, where the learning system gains a point for each correct action and loses a point for an erroneous action. The system achieves by itself by analyzing these points, and improves its performance. The basic intention of reinforcement learning is to interact with the environment and explore the job to be done properly. The objective of the system is to score the most points, and thus improve its performance. An appropriate example could be a robot which learns by itself without human intervention by moving its arms, sensing smells, and acknowledging human reactions.

3.4 Machine Learning in the Modern Era of Computing

In the modern era of computational intelligence, machine learning is achieving greater progression in research and development, and it exists everywhere around us in various modern systematic devices, such as self-driving vehicles, voice assistants, chatbots, rule-based complementary systems, and many more. It includes supervised, unsupervised, and reinforcement learning, with clustering, classification, decision tree, support vector machine algorithms, etc. Modern machine-learning models can be materialized to make predictions ranging from weather forecasting, crop yield estimating, and disease prognosis to stock market enquiries, etc. [2].

3.5 Application of Machine Learning and Its Relation to Other Fields

Artificial intelligence is viewed as an interdisciplinary field that finds its way into numerical disciplines like insights, data hypothesis, game hypothesis, and improvement. Artificial intelligence can change insight into aptitudes and can recognize authoritative examples, which is an exceptionally complicated task for humans. Unlike traditional AI, machine learning is not trying to produce an automated imitation of intelligent behaviour, but rather it tries to make use of the strengths and abilities of a powerful machine to accomplish tasks that are highly challenging for humans [1].

Recent technological innovations like machine learning and deep learning are used for data processing in various fields such as health, cyber-security, biological fields, healthcare, and food to overcome composite research challenges. Detailed learning algorithms make machine learning more specific and efficient. By using machine learning we can reduce the demand from machine-learning experts and automate processes with greater precision.

As an example, a machine-learning program can scan and process huge databases to identify patterns that are not within the scope of human observation. The understanding, or training, module in machine learning often denotes data that are arbitrarily generated. The major objective of the human is to work on those randomly generated patterns to derive meaningful conclusions that accept the concept from which these examples are taken. Machine learning has a close association with statistics, the two fields having the same objectives and methodology. There are some notable differences. If a doctor says that there is a close relationship between smoking and heart disease, it is the role of statisticians to analyze the patient samples and check the validity of the doctor's assertion. But machine learning tries to make use of the data collected from patient samples to explain the causes of heart disease. Machine learning is all about computer learning by execution; hence algorithmic issues are important. We try to improve the efficiency of the algorithm so that the learning tasks can be performed more accurately [1].

There are some additional contrasts between these two methodologies. In insights, we work with the direction of certain pre-characterized information models. AI can be effectively used in "conveyance-free" circumstances where students can discover models that provide measurable information.

3.5.1 Applying Machine Learning to Agriculture

Agriculture is the backbone of our country's economy and constitutes a major share of the world economy. In most countries, it is the principal basis for employment [3]. Countries like India still use the old ways of farming; farmers lack the knowledge and are unwilling to implement available modern technologies. The lack of knowledge about soil types, yield rate, crop varieties, weather forecasting, irrigation, improper use of pesticides, incorrect harvesting, and lack of market information has increased farmers' losses or added unexpected costs. Incorrect analysis in every segment of agriculture creates a problem or raises further issues. Unpredictable population growth increases the pressure on agricultural sectors to produce more crops as required.

There are risks in every agricultural process, from crop selection through to harvesting. Rigorous tracking of detailed crop, environment, and marketplace information helps farmers to make appropriate decisions and can reduce the risks. Technologies such as the internet of things, machine learning, deep learning, and cloud computing can be used to gather and process the details. The application of machine learning and IoT will increase production, improve quality, and increase profits for farmers and allied fields. Proper education in agriculture is very important for improving overall crop performance [4].

The following steps are executed by farmers:

Step 1: Selecting a crop

Step 2: Preparing the land for cultivation

Step 3: Sowing the seeds

Step 4: Irrigation

Step 5: Maintaining the crops

Step 6: Fertilizing

Step 7: Harvesting

Step 8: Post-harvesting activities

Figure 3.2 illustrates the sub-sectors of agricultural work.

Before harvesting, farmers perform crop selection, land preparation, seeding, irrigation, crop care and fertilization. Yield estimation is carried out to forecast production and make the necessary arrangements for harvesting or post-harvesting. Farmers also focus on other parameters such as crop or fruit maturity, market requirements, and quality. Figure 3.3 illustrates the significant characteristics of each phase of the agricultural operation.

FIGURE 3.2
Categorization of agricultural tasks.

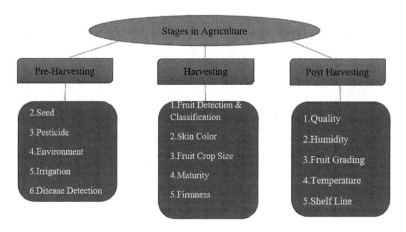

FIGURE 3.3
Stages and important factors in farming.

The most important divisions of agriculture are horticulture, cultivation, forestry, cattle, fishing, agricultural engineering, and finance. The following sections present a case study on the usage of ML in horticulture, with particular reference to fruit. The latest techniques of industrial vision systems used to classify and detect objects at each stage of farming operations are reviewed. Uses of machine learning in the pre-harvesting stage and post-harvesting stage are described. Finally, a brief overview of artificial intelligence (AI), ML and profound learning is presented [4].

3.5.1.1 Pre-Harvesting

Pre-harvest parameters are vital factors in overall crop and fruit growth. In the pre-harvesting season, ML is used to capture data such as soil parameters, seed superiority, trimming, hereditary and ecological conditions, and irrigation. It is important to focus on each element to minimize overall output losses. Here, we look at some of the chief components in pre-harvesting through which ML and ANN are used to acquire the parameters for each module [5].

3.5.1.2 Soil

In [6, 7], a soil management survey is presented in which the machine-learning techniques are applied to gather information about soil properties. Categorizing and estimating soil characteristics helps farmers to reduce the additional cost of fertilizer, reduces the demand for soil analytical experts, increases profitability, and improves soil health. Using IoT sensors makes it possible to determine parameters such as soil penetration, pH, temperature, and soil humidity. The data gathered can be given as an input value for developing a standard machine-learning model.

3.5.1.3 Seeds

Seed germination is an essential factor in seed quality, which is a significant determinant of yield and product quality. The seed germination ratio is calculated manually which is time-consuming and may lead to error. Using different ML and image-processing techniques, different automation processes can be performed for the sorting and calculation of seeds [8, 9].

3.5.1.4 Identification of Pesticides and Diseases

Prompt detection of disease is the most imperative process in keeping crops alive. Most farmers apply pesticides to crops as well. Some regularly analyze the leaves or branches of the trees and recognize diseases. These activities are based on human knowledge, which is prone to error and risk. Choice of pesticide and time of application is entirely dependent on the type of disease and its stage. The random application of pesticides to all crops can cause harm not only to crops but also to farmers. Precision agriculture helps farmers to apply the right pesticide at the right time and in the right location. Numerous studies have combined pesticide prediction with disease detection on plants [7]. Machine-learning models for disease detection and cataloguing can be established; there are novel approaches to classifying plant diseases such as deep learning. Algorithms like Plant Disease Net, Google Net CNN, ResNet50 model and SVM classifier, Alex Net Precursor and FRADD can be implemented to increase accuracy. CNN models and a fuzzy rule-based approach can also be adapted to fine-tune datasets to classify different types of diseases associated with plants and crops [10–15].

3.5.1.5 Harvesting

Once the fruits and vegetables are ready, harvesting is the most significant step. The vital parameters at this stage are fruit and crop size, skin colour, firmness, taste, quality, maturity, marketing period, fruit detection and classification for harvest. Profit is directly related to quality of crop. In the survey, we found that self-harvesting robots, machine-learning and deep-learning techniques improve outcomes and help reduce losses during the harvest stage. Several deep neural network models with different architectures can be proposed for classifying fruits/vegetables and cultures. These models can be constructed with a pre-trained DL model using a precise visual. Nevertheless, a more precise machine vision system may be proposed to classify the fruit images according to their different settings. We can renovate learning from relevant CNN models to paradigm classification models to establish maturity stage and type of fruit and vegetables, and whether they are harvestable or not [16–18].

3.5.1.6 Post-Harvest

Post-harvest is the final and critical field of agriculture that requires more consideration. After passing through all the steps from estimating the yield to harvest and post-harvest, negligence can spoil all the farmers' effort and may cause serious losses. The sub-tasks that can be taken into account at this point are: (a) growth life of fruits and vegetables; (b) post-harvest classification; and (c) export [19, 20]. A study has shown that poor handling methods after harvest may affect the fruit quality and harvest size, which may also increase overall losses. Other practices adding to losses include poor harvesting, careless handling, and poor packing and transportation conditions. Mismanagement of diseases during the production period causes disintegration with a high level of infections before harvest. Caries in the form of anthracnosis and terminal rot are very common symptoms. A DL-based classifier using Python open computer vision and end-to-end open-source platforms for ML can be used for classification, and achieves greater precision. It is also possible to propose an automatic vision system for classification after harvest [21–23].

3.5.1.7 Vital Parameters to be Considered for an Effective Agricultural Process

There are many other vital parameters, and a data-acquisition process may be required in agriculture. In addition to the details discussed in the previous section, we must also give more importance to several other parameters:

- Weather data
- Rainfall data
- Yield data

With the help of several machine-learning models and tools, we can predict the weather and rainfall levels that are desirable for the crop to grow healthily without affecting crop roots and stability. A novel multiple linear regression (MLR) model can be introduced to predict exact rainfall. The MLR model can be verified using a set of meteorological data involving monthly rainfall details in several places around the globe, which can produce better rainfall prediction performance under several performance measures.

Artificial intelligence (AI) and artificial neural networks (ANNs) are the most prominent models capable of measuring rainfall, flooding, storms, etc. An ANN model helps to detect the extraneous aspects from a dataset consisting of warmth, dampness, perspiration, and airspeed data, etc. Some types of combined models are encompassed with the fine-tuning network as well as the recurrent artificial neural network that can be well thought-out for rainfall and weather prediction.

For result analysis as well as for performance analysis, many vital measures can be considered: correlation count, mean absolute error, root mean squared error, relative absolute error and root relative squared error. These measures include diverse ML models, especially K-star, logistic regression, bagging, multilayer perceptron, radial basis function network, additive regression and Gaussian process, which can be implemented using WEKA tools [24, 25].

3.5.2 Application of Machine Learning in a Smart HealthCare System for the Elderly in Pandemic Conditions

A pandemic affects people worldwide without regard to borders. The current scenario has introduced stringent restrictions on life beyond the walls of our home. General and preventive health care is not possible during pandemic conditions, especially for the elderly. Hence here we propose a smart healthcare system that helps the elderly to obtain basic health care.

3.5.2.1 Smart Healthcare System Architecture

A living environment that contains many sensors provides an observation framework alongside an organization. It allows individual sensors to be linked to other, different sensors and to the observation frame [26]. Innovation in systems management allows a medical services framework to partner with applications from the outside world through a huge organization. Smart medical services in assisted living environments should take many aspects into consideration during planning [27]. A flexible architecture that can run on several platforms is required. The proposed smart healthcare system is made up of three layers:

1. Layer 1: Sensors
2. Layer 2: Sensor clustering into the network
3. Layer 3: Application interoperability and interconnectivity.

3.5.2.1.1 Layer 1

The first layer of the smart healthcare system is established with various sensors that sense the activities taking place where elderly people are permanently resident.

3.5.2.1.2 Layer 2

In the second layer, sensors are grouped, clustered and integrated into a single network in the smart healthcare system.

3.5.2.1.3 Layer 3

The architecture of the third layer ensures the interoperability and interconnectedness of various external domestic applications. This layer communicates to the domestic application through data networks such as mobile networks and wireless communication.

The proposed smart medical service framework gathers information, sorts and stores it. The rate at which it checks for new information and the productive correspondence convention sets up an association and sends gathered information to the information source.

The correspondence conventions are responsible for determining an effective choice and sending the detected information as information bundles. The proficiency of a smart medical services observation framework relies upon the kind of sensor and correspondence convention utilized in that organization. Since sensors in acute medical care gather an enormous amount of information, they clearly use more energy to gather and communicate information [28]. An energy-saving routing protocol is therefore required to transmit data in a sensor network specially designed for smart homes.

3.5.2.2 Proposed Smart Healthcare System

The smart house network contains various wireless sensor nodes; each node within a network constitutes a low-power transmitter. Nodes are household devices that use sensors to collect data by detecting environmental conditions [29]. The nodes in the sensor network can self-organize as a cluster and select the cluster head for efficient communication. The cluster leader spends intelligent power consumption using Leach communication protocol technology to collect and send data from all the smart home member nodes in a cluster. The limbs are the nodes of the bunch rather than the heads of the bunch (Figure 3.4).

The smart healthcare system consists of two phases: set-up and steady state.

3.5.2.2.1 Phase I

The first phase (see Figure 3.5) is the set-up phase in that all the nodes in the sensor network play the role of cluster head or member. Once their status is identified, it is announced at the node or awaits announcement of the cluster head. Thus, all nodes identify their cluster head and send an incoming request message to it [30]. At the end of the first stage, all nodes from different clusters, with the cluster head, will be ready for data transmission.

3.5.2.2.2 Phase II

The second phase (Figure 3.5) is a steady-state phase that takes on responsibility for sending data to the base station for the computational process.

In the above two phases, the smart healthcare system collects the data from the elderly person's home and provides assistance based on the data sent and received without any human-to-human physical contact.

3.6 An Overview of Artificial Intelligence and Deep Learning

The term "artificial intelligence" was first established by John McCarthy, an American computer specialist, in the year 1956. Artificial intelligence is the science and engineering method of constructing perceptive machines, specifically intelligent computer programs [31]. The term "machine learning" was first coined in the year 1959 by an American

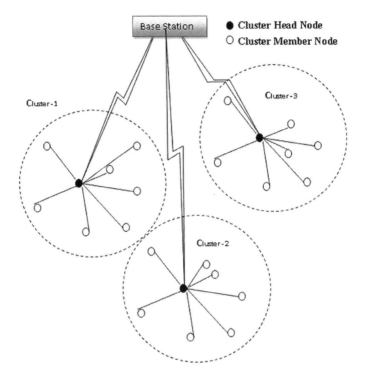

FIGURE 3.4
Smart healthcare wireless sensor communication hierarchy.

innovator, Arthur Samuel. Machine learning provides a computer with the capability to learn without being explicitly programmed, meaning ML algorithms can spontaneously learn from their knowledge. The steps involved are data collection, feature extraction, creation of models based on available data, and the choice of the cost function, optimization and adjustment of the hyperparameters, training, testing, and deployment. Deep learning (DL) is the subset of ML which helps to improve the accuracy of ML algorithms. DL takes data from higher levels (i.e., from ML) and processes it. DL technology imitates a human brain's neural network (NN). It is called DL because it works on ANNs, which consists of various layers of NN that increase with time in the training phase [32].

3.6.1 Artificial Intelligence

Artificial intelligence is one of the recent areas of computing advance that provides power to an expert system that can think and decide like human intelligence. Artificial intelligence means a human-made reasoning capacity. The artificial intelligence system may not be pre-programmed, but can use most of the powerful algorithms that can perform a task with its own intellect. It includes machine-learning algorithms such as the enhancement learning algorithm and deep learning neural networks.

3.6.2 Deep Learning

Deep learning, a subclass of ML, is one of the fastest-growing branches of AI and plays an important role in AI applications. Application of DL methods dramatically enhances the results in computer vision, image, speech, video and audio processing, object detection,

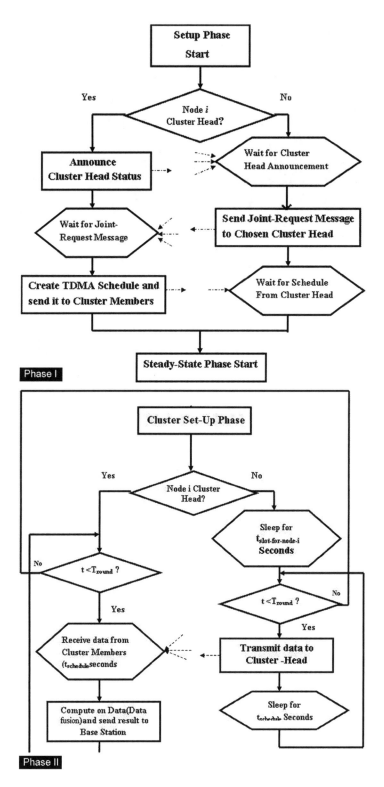

FIGURE 3.5
Smart health care phases.

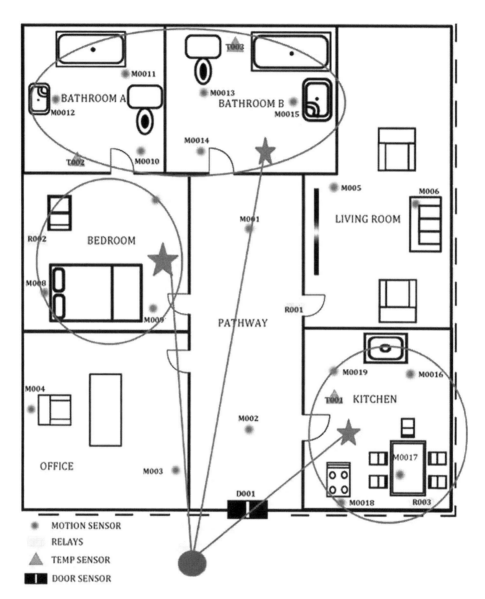

FIGURE 3.6
Smart healthcare monitoring process.

and object classification [33]. The basic components of DL are neural networks consisting of multiple layers, which gradually extract features in each layer from raw input. Some of the most prominent DL architectures are deep neural network, convolutional neural network (CNN), long short-term memory/gated recurrent unit network, deep belief network, recurrent neural network (RNN), auto-encoder, restricted Boltzmann machine, deep stacking network, and generative adversarial networks [34]. Of these, CNN and RNN are the two basic and most commonly used approaches. The CNN model consists of convolution layers, pooling layers, and activation functions which are used to build an effective model.

The performance of the CNN models depends upon the size of the labelled input dataset, the number of layers used in the architecture, and the duration of training [35].

Building a new model from scratch is time-consuming. Moreover, high-end hardware support in the dataset is vital for building a highly accurate model. The cleaner and more equally distributed the input dataset, the greater accuracy you will achieve. To overcome the problems of time required for training, high-end hardware requirement, and a large dataset, "transfer learning", is used. Transfer learning helps to build accurate models in less time and has shown remarkable results in image classification [36].

Transfer learning can be defined as follows:

> If there is an origin of basis sphere and a training task, a target domain and a learning task, transfer learning tries to improvise the learning of the target predictive function into learning task using the knowledge acquired from the source and target domain [37].

The CNN model layers are divided into two parts: the extraction layers of the characteristics and the classification layers. Entity extraction layers are also called convolution layers: these are composed of convolution layers and grouping layers. The classification layers are made of fully interconnected layers. Transfer learning can be applied in the following scenarios: (a) train the entire model: use the pre-trained model and train it on its dataset; (b) fixed convolution layers: take a pre-trained model, freeze the feature-extraction layers and modify the classification layers as per your need; and (c) fine-tune the ConvNet: use the pre-trained model and freeze a few convolution layers instead of all of them and retrain the model on your dataset [37].

3.7 Conclusion

This chapter aims to provide a healthier awareness of some of the implementations and uses of machine-learning algorithms: artificial intelligence, deep learning, CNN, RNN, wireless sensor networks, IoT and several other methods in various fields that are vital for real-world scenarios. Machine learning can ultimately provide superior use of technology, implementation, and supplementary mathematical as well as statistical ideas to move our world on to a better place for succeeding generations to live a healthy and peaceful life.

References

1. Shai Shalev Shwartz & Sai Ben David, *Understanding Machine Learning From Theory to Algorithm*, Cambridge University Press, 2014.
2. Machine Learning, n.d. javapoint.com
3. The World Factbook— Central Intelligence Agency. www.CIA.gov. Retrieved, 2017.
4. Vishal Meashram, Applications of Machine Learning in Agriculture Domain—A State of Art", *International Journal of Advanced Science and Technology*, 29(8), 5319–5343 (2020).
5. I.K. Arah, H. Amaglo, E.K. Kumah, and H. Ofori, Preharvest and Postharvest Factors Affecting the Quality and Shelf Life of Harvested Tomatoes: A Mini-Review, International Journal of Agronomy, (2015), Article ID 478041.

6. N. Gnanasankaran, and E. Ramaraj, An Effective Yield of Paddy crop in Sivaganga district - An Initiative for Smart Farming, *International Journal of Scientific and Technology Research*, 9(2), 6553–6556 (2020).

7. D. Sivakumar. et al. Computerized Growth Analysis of Seeds Using Deep Learning Method, 7(6S5), 1885–1892 (2019).

8. Eng Huang et al., Research on Classification Method of Maize Seed Defect Based on Machine Vision, *Hindawi Journal of Sensors*, (2019), Article ID 2716975.

9. Susu Zhu et al., Near-Infrared Hyperspectral Imaging Combined with Deep Learning to Identify Cotton Seed Varieties, 24(18), 3268 (2019).

10. M. Hammad Saleem, Review Plant Disease Detection and Classification by Deep Learning, *Plants*, 8(11), 468 (2019).

11. R. K. Prange, Pre-Harvest, Harvest and Post-Harvest Strategies for Organic Production of Fruits and Vegetables, *Acta Hortic*, 933, 43–50 (2012).

12. M. Hammad Saleem, Review Plant Disease Detection and Classification by Deep Learning, *Plants*, 8(11), 468 (2019).

13. Muammar Turkoglu, Plant Disease and Pest Detection Using Deep Learning-Based Features, *Turkish Journal of Electrical Engineering & Computer Science*, 27, 1636–1651 (2019).

14. Bin Liu, Identification of Apple Leaf Diseases Based on Deep Convolutional Neural Networks, *Symmetry*, 10(1), 11 (2018).

15. Vippon Preet Kour et al., *Fruit Disease Detection Using Rule-Based Classification*, Springer Nature Singapore Pte Ltd, 295–312 (2019).

16. Shuli Xing, Citrus Pests and Diseases Recognition Model Using Weakly Dense Connected Convolution Network, *Sensors*. 19(14), 3195 (2019).

17. Benjamin Doh et al., Automatic Citrus Fruit Disease Detection by Phenotyping Using Machine Learning, *International Conference on Automation & Computing*, Lancaster University, Lancaster, UK, 5–7 September 2019.

18. K. Bresilla, G. D. Perulli, A. Boini, B. Morandi, L. Corelli Grappadelli, and L. Manfrini, Single-Shot Convolution Neural Networks for Real-Time Fruit Detection within the Tree. *Front, Plant Science Journal*, 10, article 611 (2016).

19. M. Shamim Hossain, and Muneer Al-Hammadi, and Ghulam Muhammad, Automatic Fruits Classification Using Deep Learning for Industrial Applications, *IEEE Transactions on Industrial Informatics*, 15(2), 1027–1034 (2019).

20. Raymond Kirk et al. L*a*b*Fruits: A Rapid and Robust Outdoor Fruit Detection System Combining Bio-Inspired Features with One-Stage Deep Learning Networks, *Sensors*, 20(1), 275 (2020).

21. Rosanna Ucat, and Jennifer Cruz, Postharvest Grading Classification of Cavendish Banana Using Deep Learning and Tensor Flow, *ISMAC*, 2019.

22. David Ireri, A Computer Vision System for Defect Discrimination and Grading in Tomatoes Using Machine Learning and Image Processing, *Artificial Intelligence in Agriculture*, 22, 28–37 (2019).

23. Piedad Eduardo Jr, Postharvest Classification of Banana (Musa Acuminata) Using Tier-Based Machine Learning, *Postharvest Biology and Technology*, 145, 93–100 (2018).

24. N. Gnanasankaran, and E. Ramaraj, A Multiple Linear Regression Model to Predict Rainfall Using Indian Meteorological Data, *International Journal of Advanced Science and Technology*, 29(8), 746–758 (2020).

25. N. Gnanasankaran, and E. Ramaraj, An Intelligent Framework for Rice Yield Prediction Ussing Machine Learning Based Models, *International Journal of Scientific and Engineering Research*, 12(1), 422–431 (2021).

26. G. Rakesh, and K. Thangadurai, Optimistic Algorithmic Approaches for Traffic Engineering Policies of Congestion and Traffic Distribution in the Networks, *International Journal of Computer Applications*, 173(1), 10–14 (2017).

27. G. Rakesh, and K. Thangadurai, TDA Routing for Internet Backbone to Ensure Optimized Computing in Networks, *International Journal of Computer Science & Networks*, 16(2), 42–45 (2016).

28. G. Rakesh, and K. Thangadurai, "Congestion Identification and Avoidance Approach to Internetworking for Traffic Engineering Mechanism, *International Journal of Advanced Research in Computer Engineering & Technology*, 6(12), 1867–1872 (2017).
29. G. Rakesh, Building an Optimized END TO END Automated and Secured Solution Using Both Sequences through the Internet of Things, *International Conference on Humanities, Technology and Science*, Asia Pacific University, Malaysia, 2018.
30. G. Rakesh, Hybridized Gradient Descent Spectral Graph and Local—Global Louvain Based Clustering of Temporal Relational Data, *International Journal of Engineering & Advanced Technology*, 9(3), 3515–3521 (2020).
31. Md Zahangir Alom, A State-of-the-Art Survey on Deep Learsning Theory and Architectures, *Electronics*, 1, 1–67 (2019).
32. Shaveta Dargan, A Survey of Deep Learning and Its Applications: A New Paradigm to Machine Learning, *Archives of Computational Methods in Engineering, Springer*, 45–84 (2019).
33. Yann Le Cun et al., Deep Learning, *Nature*, 152–212 (2018).
34. Samira Pouyanfar, Survey on Deep Learning: Algorithms, Techniques, and Applications, *ACM Computing Surveys*, 652–698 (2018).
35. Mi Lu, A Probe Towards Understanding GAN and VAE Models, arXiv, 541–549 (2018).
36. https://cs231n.github.io/transfer-learning
37. https://towardsdatascience.com/transfer-learning-from-pre-trained-models
38. G. Rakesh, and K. Thangadurai, A Short Study on Routing and Traffic Engineering Design for Global Internetworking, *IJARCSSE*, 4(2), 240–248 (2014).
39. G. Rakesh, and K. Thangadurai, Traffic Stationed Routing Algorithm for Traffic Engineering to Ensure Quality of Service in Internetworking Operations, *European Journal of Scientific Research*, 134(4), 355–361 (2015).

4

Prediction of Diabetics in the Early Stages Using Machine-Learning Tools and Microsoft Azure AI Services

Chandrashekhar Kumbhar and Abid Hussain

Career Point University, Kota, India

CONTENTS

4.1 Introduction

There are four kinds of diabetes, types 1 and 2 being the most common. There is no cure for type 1, which occurs at a relatively young age, generally before the age of 35, and is very difficult to detect. Patients with type 2 diabetes who develop it in their mid-to-late adult years can be treated with medication and regular exercise (Swapna et al., 2018). Research using the PIMA Indian diabetes dataset has been used to forecast diabetes.

This study uses a real-time dataset to predict diabetes (Roychowdhury et al., 2013). A licenced copy of the RapidMiner tool was used, which is able to quickly visualize the data and apply numerous operations/methods and algorithms on it (Sharma and Singh, 2018). In RapidMiner, we employed all of the supporting algorithms, but this chapter describes two supervised machine-learning algorithms—K Nearest Neighbour and Random Forest—that have achieved accuracy of more than 89% (Kaur, 2020).

The rest of the chapter is organized as follows. Section 4.2 gives detailed information about the dataset and Section 4.3 describes the tools used. Section 4.4 describes data pre-processing. Section 4.5 provides visualization and analysis of the preprocessed data. Sections 4.6 and 4.7 explain when applying KNN and Random Forest, respectively, through RapidMiner are appropriate. Section 4.8 describes the application of KNN, Random Forest and decision tree through Microsoft Azure AI Services. In Section 4.9, a comparative analysis of both prediction models is provided.

4.1.1 Risk Factors for Diabetes

4.1.1.1 Type 1

- Family history
- Specific genetic characteristics
- Conditions such as cystic fibrosis or haemochromatosis

4.1.1.2 Type 2

- Some medical disorders, such as cystic fibrosis or haemochromatosis, are helped by a specific weight loss approach.
- Being overweight
- Smoking
- Unhealthy diet
- Gestational diabetes

4.1.1.3 Pre-Diabetics

- Overweight
- Over the age of 45
- People with a family history of type 2 diabetes
- Have had gestational diabetes

4.1.1.4 Gestational Diabetes

- Diagnosed with gestational diabetes in a previous pregnancy
- Polycystic ovary syndrome (PCOS)
- Family history of type 2 diabetes

4.2 Dataset Collection

Dataset collection was done using Google Forms, with additional data for educational purposes obtained from a pathology lab (Kopitar et al., 2020). The attributes particularly studied are HBA1C test details, family history, and symptoms. Figure 4.1 is a screenshot of the dataset.

4.3 Tools Used for Prediction

4.3.1 Orange

Orange is an open-source data visualization and machine-learning application that is used to create visual data analysis processes enabling the required data to be seen clearly (Rubaiat et al., 2018).

4.3.2 RapidMiner

RapidMiner is a data science platform that can perform essential tasks including data preparation, data visualization, machine learning, data analysis and deep learning. It includes a number of modelling, feature selection, and extraction options. Some machine-learning techniques also apply to these results (Haq et al., 2020).

1	Diabtetes	HBA1C	Gender	Family membe	Type	Sam 1	Sam 2	Insulin	Issues	Health problem	Weight changes
2	Yes	1	Female	1	1	Frequent urination	Extreme hunger	1	Afraid, Upset	High Blood Pressure	Yes
3	No	0	Male	1	0	Frequent urination	Extreme hunger	0	Confused, Upset, Feel like I canâ€™t do	No I do not have	No
4	No	0	Male	0	0	Frequent urination	Unintended weight loss	0	Afraid	No I do not have	No
5	Yes	1	Male	1	2	Extreme hunger	Unintended weight loss	1	Angry, Afraid, Sad	Heart Disease	Yes
6	Yes	1	Female	1	2	Fatigue and weakness	Feeling very tired	1	Upset	Thyroid	No
7	Yes	1	Female	1	2	Extreme hunger	Blurred vision	1	Angry	High Blood Pressure	Yes
8	Yes	0	Female	1	1	Extreme hunger	Fatigue and weakness	1	Upset, Feel like I canâ€™t do my job	No I do not have	Yes
9	Yes	1	Male	1	1	Extreme hunger	Unintended weight loss	1	Angry, Sad, Feel like I am a sick person	No I do not have	Yes
10	No	1	Male	0	0	Extreme hunger	Unintended weight loss	0	ngry, Sad, Feel like I canâ€™t do my jol	High Blood Pressure	Yes
11	Yes	1	Male	1	1	tability and other mood chan	Fatigue and weakness	1	ngry, Sad, Feel like I canâ€™t do my jol	High Cholesterol/Trigh	Yes
12	No	0	Male	0	0	Blurred vision	Feeling very tired	0	canâ€™t live the way I want, Feel like I	High Blood Pressure, A	No
13	No	0	Female	0	0	Extreme hunger	Irritability and other mood changes	0	Afraid, Confused, Sad	No I do not have	No
14	Yes	1	Male	1	1	Blurred vision	Feeling very tired	1	ad, Upset, Feel like I canâ€™t do my jol	High Blood Pressure	Yes
15	Yes	1	Male	1	2	Extreme hunger	Unintended weight loss	1	Angry, Afraid, Sad	Heart Disease	Yes
16	No	0	Female	0	0	Fatigue and weakness	Feeling very tired	0	Upset	Thyroid	No
17	Yes	1	Female	1	2	Extreme hunger	Blurred vision	1	Angry	High Blood Pressure	Yes
18	No	1	Female	1	0	Extreme hunger	Fatigue and weakness	0	Upset, Feel like I canâ€™t do my job	No I do not have	No
19	Yes	1	Male	1	1	tability and other mood chan	Fatigue and weakness	1	ngry, Sad, Feel like I canâ€™t do my jol	High Cholesterol/Trigh	Yes
20	No	1	Male	1	0	Blurred vision	Feeling very tired	0	canâ€™t live the way I want, Feel like I	High Blood Pressure, A	No
21	Yes	1	Female	0	2	Extreme hunger	Irritability and other mood changes	1	Afraid, Confused, Sad	No I do not have	Yes
22	Yes	1	Male	1	2	Blurred vision	Feeling very tired	1	ad, Upset, Feel like I canâ€™t do my jol	High Blood Pressure	Yes
[illegible]	[illegible]	[illegible]	[illegible]	[illegible]	[illegible]	[illegible]	[illegible]	[illegible]	[illegible]	[illegible]	[illegible]

FIGURE 4.1
Dataset.

4.3.3 Microsoft Azure

Microsoft Azure is a cloud foundation platform for creating, managing, and deploying different services in the cloud. Many Azure AI services are accessible free, while others are only available for purchase. Microsoft Azure includes tools to help with machine learning and data visualization.

4.4 Data Cleansing

When the collected Google Forms and pathology lab data were combined, the data were found to be inadequate. It was therefore decided to clean the dataset using the RapidMiner tool. RapidMiner provides more than 300 operators that can be used to preprocess the dataset, apply models, obtain results, and perform validation (Indoria and Rathore, 2018). Tasks in data cleaning include, for example, normalization, binning, missing, duplicate, outliers and reduction of dimensionalities.

4.4.1 Normalization

The goal of normalization is to change the values of numeric columns in the dataset to use a common scale (Yahyaoui et al., 2019). The three methods available in RapidMiner are: normalizing, de-normalizing, and scaling by weight.

The input attributes vary according to diabetes type and insulin, and there is a range of HbA1C values. Normalization was performed using Z-transform method (Figure 4.2).

4.4.2 Missing Data

In machine learning, data is said to be king, meaning that if you have good data, you'll achieve good accuracy. But if the data are not good, you can make them good by using

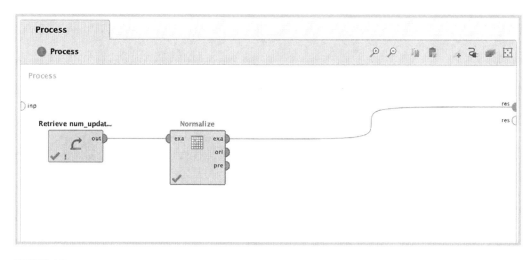

FIGURE 4.2
Normalization using Z- transformation technique.

Missing operators from RapidMiner. The eight available operator types are (Chetoui, 2018) (Figures 4.3–4.5):

1. Replace Missing Values
2. Impute Missing Values
3. Declare Missing Values
4. Replace Infinite Values

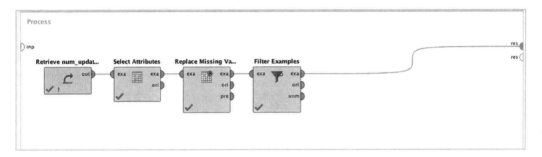

FIGURE 4.3
Replace Missing Values operator.

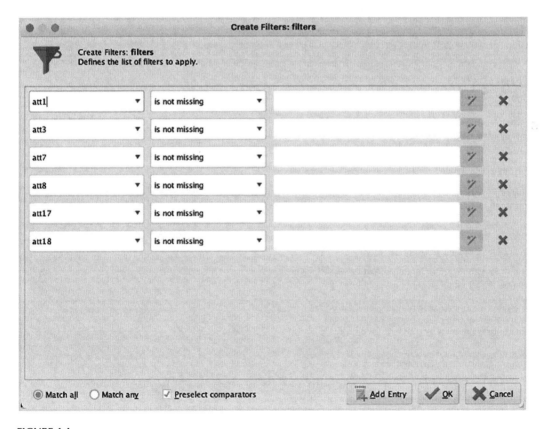

FIGURE 4.4
Removing missing values.

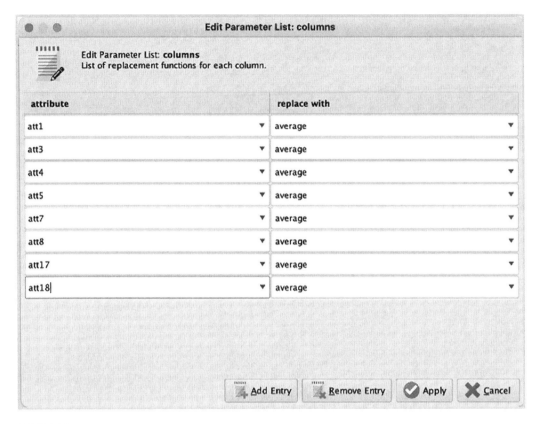

FIGURE 4.5
Applying filters.

5. Remove Unused Values
6. Fill Data Gaps
7. Replace All Missings
8. Handle Unknown Values

Data were collected from diabetic and non-diabetic patients displaying some sort of symptoms. In a few cases or attributes, values are missing, so the Replace Missing Values operator was used to fill these (Wei et al., 2018).

4.5 Dataset Visualization

Data visualization is the representation of data in charts and graphs that helps to understand the various types of data and how to deal with it. Generally, analysis of the data determines the type of machine-learning algorithm to be used for further processing (Samant and Agarwal, 2018).

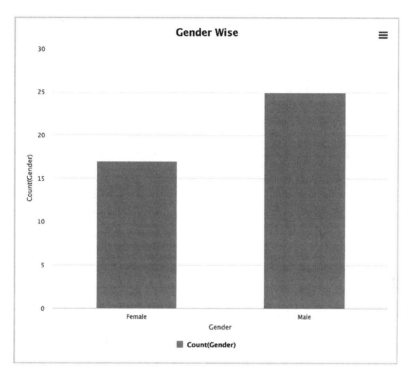

FIGURE 4.6
Gender-wise count.

4.5.1 Bar Plot (Gender)

The survey showed a slight prevalence of diabetes in males (12%) as opposed to females (11.7%) in India (Sisodia and Sisodia, 2018) (Figure 4.6 and Figure 4.7).

4.5.2 Bar Plot (HbA1c)

The HbA1c test establishes a patient's blood glucose level in the last three months with the help of red blood cells. An HbA1c range of < 5.7 is normal, HbA1c > 5.7 to 6.5 is pre-diabetic and HbA1c > 6.5 is diabetic (Islam and Jahan, 2017). When HbA1c is controlled, a patient can live a normal life (Figure 4.8). If the level is too high, symptoms include feeling angry, upset or sick.

4.5.3 3D Scatter Plot (HbA1c)

A scatter plot shows the relationship between two variables, or how two variables are correlated with each other (Kumari and Chitra, 2013). A three-dimensional scatter plot is like a scatter plot, but with three variables. In our 3D scatter plot, the x axis represents symptoms, y axis issues and z axis gender.

As mentioned for a few cases, a person may have frequent urination and extreme hunger. We found that extreme hunger has relatively high symptoms in diabetic patients (Figure 4.9).

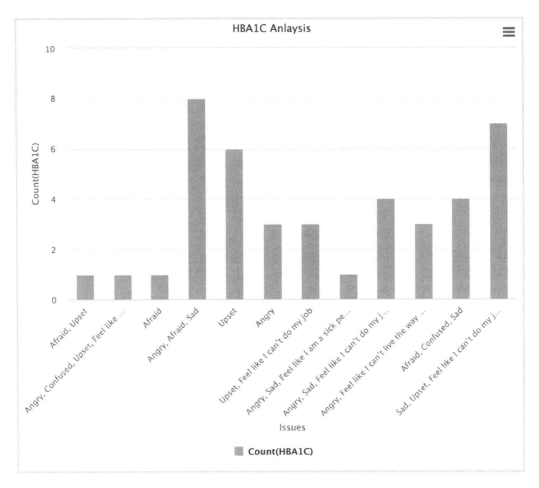

FIGURE 4.7
HbA1c count by symptoms.

4.5.4 Bell Curve

A bell curve is a type of graph that is used to visualize the distribution of a set of chosen values across a specified group that tend to have normal values in the centre, and peaks with low and high extremes tapering off relatively symmetrically on either side (Zou et al., 2018). Bell curves are visual representations of normal distribution, also called Gaussian distribution. We found that extreme hunger is a relatively high symptom in diabetic patients (Figure 4.9). We used HbA1c attribute to find the probability from 0 to 1.

4.6 KNN Implementation

A K Nearest Neighbour is a supervised machine-learning algorithm. Assuming that the new case/data and existing cases are comparable, the KNN algorithm will place the new case in the category that is most similar to the existing categories (Sowah, 2020) (Figure 4.10).

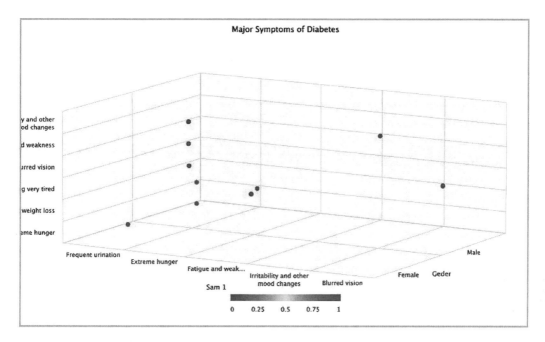

FIGURE 4.8
Major symptoms of diabetes.

Each data point in the KNN algorithm is classified according to its similarity with the existing data. When fresh data emerges, therefore, it can be readily categorized, using the KNN algorithm, into an appropriate category. No assumptions are made about the underlying data when using the KNN method. When it is time to classify the data, it acts on the stored information, which is why it is also known as a "lazy learner" method (Mujumdar and Vaidehi, 2019).

The following RapidMiner operators were used:

1. **Retrive_num_update:** Name of the dataset used for implementation.
2. **Set Role:** This operator is used to set the role of attributes, here we have used it to label the role of the target variable.
3. **Select attribute:** An optional operator where we can use all the attributes or only selected attributes as input to the algorithm.
4. **Split data:** This operator is used to split the data in training the testing phase. Here we have used 70% data for training and 30% data for testing purposes, which we have found to be the best and most accurate combination.
5. **KNN:** The KNN operator/algorithm is used to find the output. Here we have applied k = 5 to the algorithm. Next to the K parameter we have ticked the weighted vote parameter because if this parameter is set, the distance values between the examples are also taken into account for the prediction. MixedMeasure has been selected from MixedMeasures, NominalMeasures, NumericalMeasures and BregmanDivergences. Within MixedMeasure, MixedEuclideanDistance has been selected for future use.

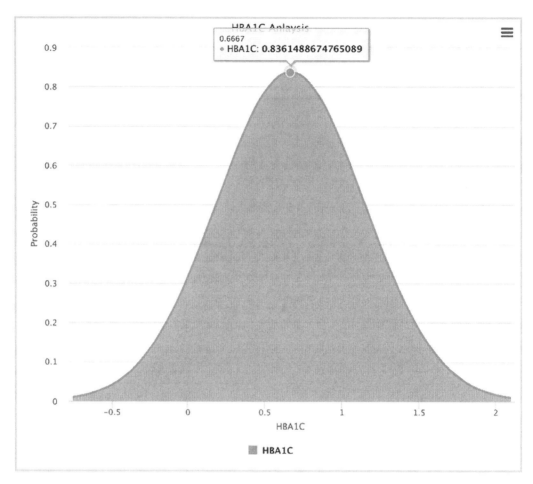

FIGURE 4.9
Bell curve probability for HBA1C.

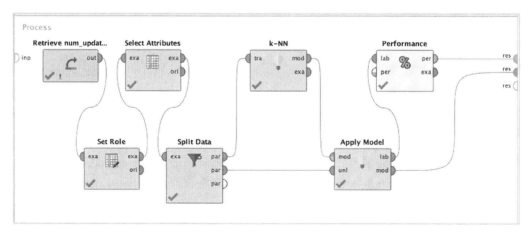

FIGURE 4.10
K nearest neighbour.

...	Diabtetes	prediction(Diabtetes)	confidence(Yes)	confidence(No)	HBA1C	Gender	F...	T...	Sam 1	Sam 2
1	No	No	0.033	0.967	0	Male	0	0	Frequent ...	Unintend...
2	Yes	Yes	0.995	0.005	1	Female	1	2	Extreme h...	Blurred vi...
3	Yes	Yes	0.983	0.017	1	Male	1	1	Extreme h...	Unintend...
4	Yes	Yes	0.966	0.034	1	Male	1	1	Irritability ...	Fatigue a...
5	Yes	Yes	0.994	0.006	1	Male	1	2	Extreme h...	Unintend...
6	No	No	0.012	0.988	0	Female	0	0	Fatigue an...	Feeling v...
7	Yes	Yes	0.995	0.005	1	Female	1	2	Extreme h...	Blurred vi...
8	No	No	0.306	0.694	1	Female	1	0	Extreme h...	Fatigue a...
9	No	No	0.262	0.738	1	Male	1	0	Blurred vi...	Feeling v...
10	Yes	Yes	0.988	0.013	1	Female	1	2	Fatigue an...	Feeling v...
11	Yes	Yes	0.994	0.006	1	Male	1	2	Extreme h...	Unintend...
12	No	No	0.068	0.932	0	Male	1	0	Irritability ...	Fatigue a...
13	No	No	0.012	0.988	0	Male	1	0	Blurred vi...	Feeling v...
14	Yes	Yes	0.978	0.022	1	Female	0	2	Extreme h...	Irritability...
15	Yes	Yes	0.807	0.193	0	Male	1	1	Blurred vi...	Feeling v...
16	Yes	Yes	0.998	0.003	1	Male	1	1	Blurred vi...	Feeling v...

FIGURE 4.11
Prediction and confidence using KNN.

6. **Apply model:** Takes the remaining 30% testing data to test.
7. **Performance:** This operator finds the performance of the algorithm or finds the confusion matrix.

Once all the above operators have been selected, we can confidently hit the "run" button to obtain a table for our target variable (Figure 4.11).

Figure 4.11 shows the prediction of the target variable. Using HbA1c and related input variables, the model is predicting whether or not the patient has diabetes. An accuracy level of more than 89% has been obtained. Future attempts will aim to increase this level (Kumar Dewangan and Agrawal, 2015).

4.7 Random Forest Implementation

Random Forest is a common machine-learning method in supervised learning that may be used for both classification and regression issues, depending on the application. Complicated problems can be tackled and the model's performance enhanced using what is known as "ensemble learning", which combines many classifiers to solve a problem (Reddy et al., 2020) (Figure 4.12).

Random Forest uses several decision trees on different subsets of a dataset and averages them to enhance prediction accuracy. Because the Random Forest does not depend on a single decision tree, it forecasts the ultimate outcome by combining predictions from several trees (Kad and Kumbhar, 2019).

Because of the increased quantity of trees in the forest, there is less overfitting.

The following RapidMiner operators were used (Chauhan, 2014) (Figures 4.13–4.19):

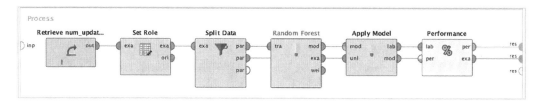

FIGURE 4.12
Random Forest implementation.

...	Diabtetes	prediction(Diabtetes)	confidence(Yes)	confidence(No)	HBA1C	Gender	F...	T...	Sam 1	Sam 2
1	No	No	0.033	0.967	0	Male	0	0	Frequent ...	Unintend...
2	Yes	Yes	0.995	0.005	1	Female	1	2	Extreme h...	Blurred vi...
3	Yes	Yes	0.983	0.017	1	Male	1	1	Extreme h...	Unintend...
4	Yes	Yes	0.966	0.034	1	Male	1	1	Irritability ...	Fatigue a...
5	Yes	Yes	0.994	0.006	1	Male	1	2	Extreme h...	Unintend...
6	No	No	0.012	0.988	0	Female	0	0	Fatigue an...	Feeling v...
7	Yes	Yes	0.995	0.005	1	Female	1	2	Extreme h...	Blurred vi...
8	No	No	0.306	0.694	1	Female	1	0	Extreme h...	Fatigue a...
9	No	No	0.262	0.738	1	Male	1	0	Blurred vi...	Feeling v...
10	Yes	Yes	0.988	0.013	1	Female	1	2	Fatigue an...	Feeling v...
11	Yes	Yes	0.994	0.006	1	Male	1	2	Extreme h...	Unintend...
12	No	No	0.068	0.932	0	Male	1	0	Irritability ...	Fatigue a...
13	No	No	0.012	0.988	0	Male	1	0	Blurred vi...	Feeling v...
14	Yes	Yes	0.978	0.022	1	Female	0	2	Extreme h...	Irritability...
15	Yes	Yes	0.807	0.193	0	Male	1	1	Blurred vi...	Feeling v...
16	Yes	Yes	0.998	0.003	1	Male	1	1	Blurred vi...	Feeling v...

FIGURE 4.13
Prediction and confidence using Random Forest.

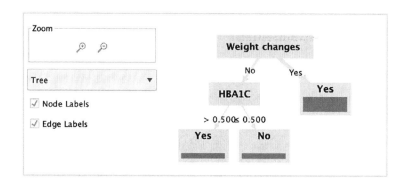

FIGURE 4.14
RF model 1.

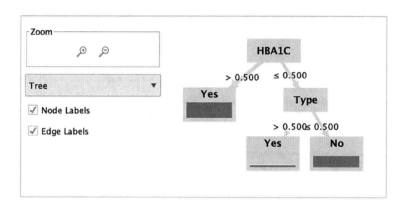

FIGURE 4.15
RF model 2.

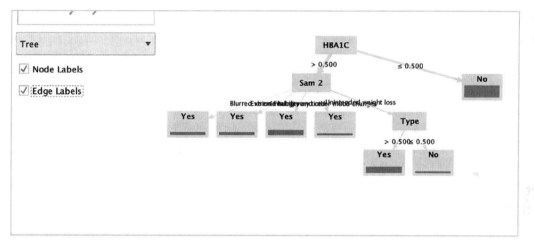

FIGURE 4.16
RF model 3.

1. **Retrive_num_update_1:** Name of the dataset used for implementation.
2. **Set Role:** This operator is used to set the role of attributes. Here it is used to set the role of the target variable as label.
3. **Split data:** This operator is used to split the data in training the testing phase. Here we have used 75% data for training and 25% data for testing purpose, which we have found to be the best and most accurate combination.
4. **Random Forest:** This is actual algorithm that was applied. The parameters set are: number of trees, 100; criterion, gain ratio; maximal depth, 1. We have the option to apply pruning and pre-pruning techniques.
5. **Apply model:** An important RapidMiner operator. Apply model takes the remaining 30% testing data to test.
6. **Performance:** This operator finds the performance of the algorithm /finds the confusion matrix.

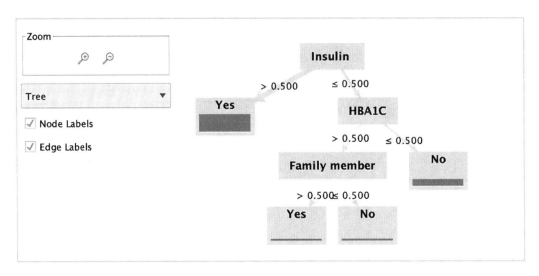

FIGURE 4.17
RF model 4.

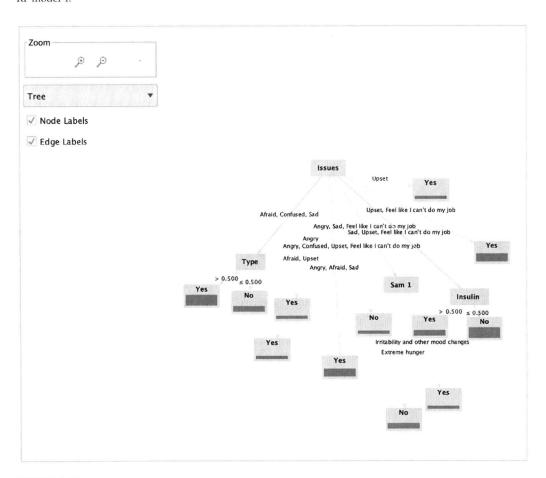

FIGURE 4.18
RF model 5.

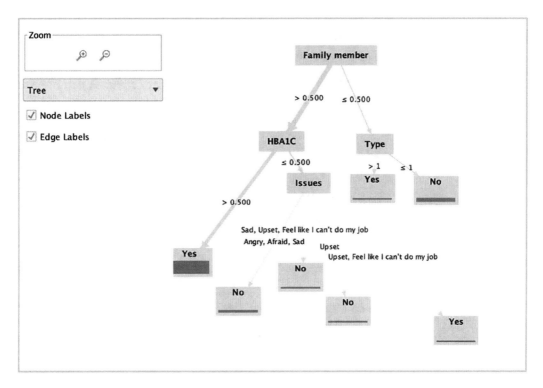

FIGURE 4.19
RF model 6.

Figure 4.14 shows an RF model that has generated a tree with weight changes as root node. A patient whose weight has changed in the last five to six months may be diabetic (Hofmann and Klinkenberg, 2016).

In Figure 4.15 HbA1c is the root. If this increases above the normal ratio, the patient may have diabetes.

Figure 4.16 shows the RF model for HbA1c attribute as a root node. If HbA1c is high, it is checked against childhood symptoms to make predictions.

In Figure 4.17, insulin is the root node. A person who is taking insulin is mostly diagnosed with type 1 diabetes.

Figure 4.18 is the most important and valuable tree in this implementation. The issues attribute is shown as a root node, and the structure or roadmap of prediction is elaborated (Azure, 2016).

Sometimes family history is significant. Figure 4.19 shows the RF model for the family member attribute as a root node. A patient with symptoms having a blood relation with diabetes has a chance of being diagnosed.

4.8 Microsoft Azure Implementation

In addition to helping data scientists enhance their productivity by automating many of the time-consuming activities associated with training models, Azure Machine Learning leverages cloud-based computational resources that scale effectively to accommodate enormous amounts of data while incurring expense only when actually needed.

The following are the implementation steps of early-stage diabetes prediction using the Azure Machine Learning cloud service (Kotu and Deshpande, 2014).

Step 1: Create Azure Machine Learning workspace.

The Microsoft Azure Machine Learning workspace can be used to manage machine-learning workloads, such as data, computing resources, code, models and other assets (Kotas et al., 2018) (Figure 4.20).

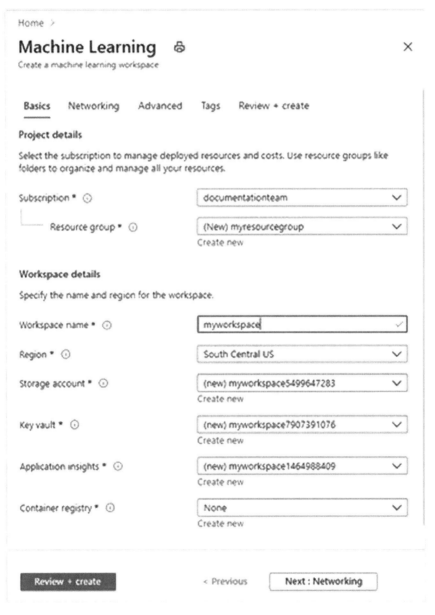

FIGURE 4.20
Creating Azure Machine Learning workspace.

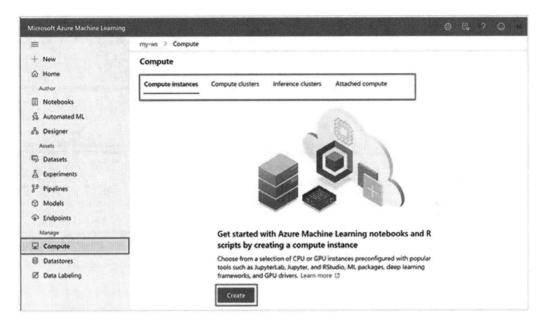

FIGURE 4.21
Create compute instance.

Step 2: Create computing resources
Once the Azure Machine Learning workspace has been built, it can be used to manage the different assets and resources required. Best results can be obtained by using the Standard DS11 v2 image (Verma et al., 2014).

It should be remembered that computing resources must be cleaned up after completing each module (Figure 4.21).

Step 3: Explore the data (create dataset)
The dataset can be uploaded from a local machine or from the Internet. It can then be visualized in various graphs or charts (Sundas, 2021).

In the diabetes dataset which was used for RapidMiner implementation, there are 11 attributes in total: diabetes, HbA1c, gender, family member already diagnosed, type, symptoms 1, symptoms 2, insulin status, issues, health problems, and weight changes in the last six months. Diabetes is set as the target variable which is in categorical form (Han et al., 2008) (Figure 4.22).

Step 4: Train a machine-learning model.
Using the scalability of cloud computing, Azure Machine Learning can automatically test several preprocessing approaches and model-training algorithms in parallel in order to identify the highest performing supervised machine-learning model for the data at hand (Mund, 2015).

Azure Machine Learning can be used to create supervised machine-learning models which use training data that includes label values that are known (Figures 4.23 and 4.24).

Models may be trained using automated machine learning in the following areas: (i) classification; (ii) regression; and (iii) time series.

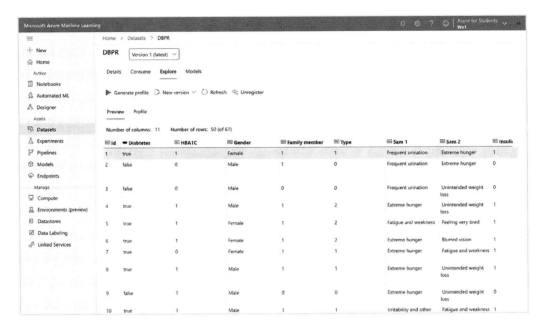

FIGURE 4.22
Dataset in Azure.

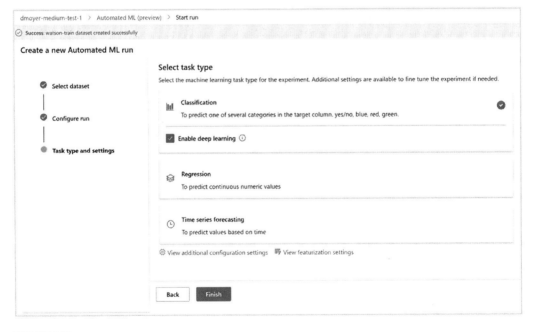

FIGURE 4.23
Select classification model.

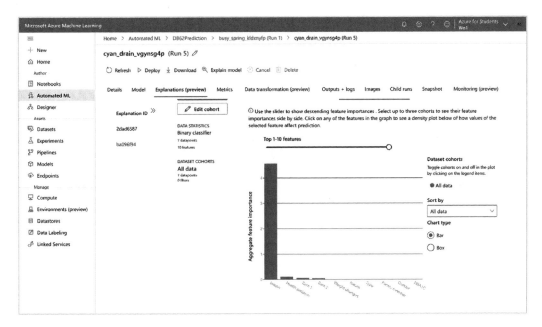

FIGURE 4.24
Visualization of parameters using aggregate function.

In this study, all other algorithms except Random Forest, decision tree, and KNN were blocked in Additional Settings. The model created can be visualized from the Explanations or Metrics tab (Ristoski et al., 2015).

Step 5: Deploy a model as service

This allows the best-performing machine-learning model to be deployed as a service for client applications.

It is possible to run Azure Machine Learning services on ACIs (Azure Container Instances) or AKS (Azure Kubernetes Service) clusters. An AKS deployment is suggested for production applications. To do this, an inference cluster computing target must be built. An ACI service is an appropriate deployment target for testing which does not require an inference cluster to be established (Barga et al., 2015).

Step 6: Test the model

There is a variety of options for testing models. The Jupyter notebook built into Microsoft Azure cloud service can be used to write test cases and test them with a training model (Wilder, 2012). Figure 4.25 shows a false result from the testing model, indicating the patient is not diagnosed as diabetic and has less chance of being so in future.

In this last step data is provided to test the model and receive the result as false. That means the person has not been diagnosed and there are less chances in future as well.

In Figure 4.26 a true result has been obtained, meaning the person is diagnosed with diabetes without any family history.

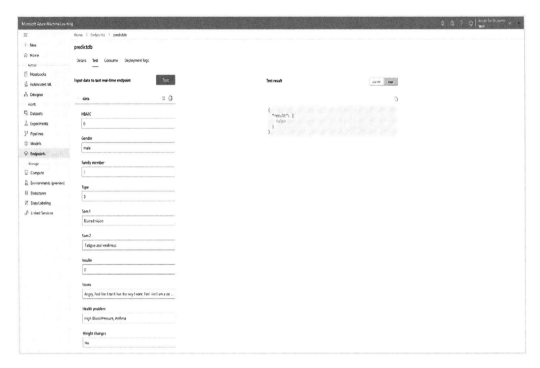

FIGURE 4.25
Testing model 1.

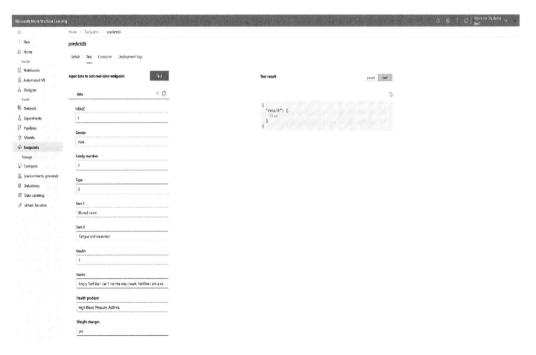

FIGURE 4.26
Testing of model 2.

TABLE 4.1

Comparison of RapidMiner and Microsoft Azure (AI Service)

Sr. No	Parameters	RapidMiner	Microsoft Azure
1	Tool used	Data science software platform	Microsoft cloud service
2	Algorithm used	KNN, Random Forest	Decision tree, KNN, Random Forest
3	Cost	Free	Paid
4	Accuracy of model	KNN = 87%, Random Forest = 85%	KNN = 89%,

4.9 Comparison of RapidMiner and Microsoft Azure

The comparison summary of the two platforms used for predictions applies only for the model implemented in the present work. We found both the models appropriate for visualization and implementation (Ertek, 2013).

4.10 Conclusion and Future Scope

In recent decades, diabetes has been spreading rapidly among the younger generation, and while we cannot stop the spread, it can be controlled with the help of machine-learning algorithms through Microsoft Azure services or the RapidMiner tool. The parameters used in these supervised learning algorithms, such as HbA1c, have never been tried before. In RapidMiner, using KNN and Random Forest algorithms, accuracy of more than 85% has been achieved, and with Microsoft Azure AI Service for KNN, 89% accuracy.

Future research will attempt to design an IoT device which can provide HbA1c results through device sensors as early as possible, so that the early stages of diabetes can be easily predicted.

References

Azure, M. (2016). *Microsoft Azure. [línea]*. Available: https://docs.microsoft.com/es-es/azure/virtual-machines/linux/quick-createportal (Last accessed December 10, 2017).

Barga, R., Fontama, V., Tok, W. H., & Cabrera-Cordon, L. (2015). *Predictive analytics with Microsoft Azure machine learning* (pp. 221–241). Berkely, CA: Apress.

Han, J., Rodriguez, J. C., & Beheshti, M. (2008, December). Diabetes data analysis and prediction model discovery using RapidMiner. In *2008 Second International Conference on Future Generation Communication and Networking* (Vol. 3, pp. 96–99). IEEE.

Haq, A. U., Li, J. P., Khan, J., Memon, M. H., Nazir, S., Ahmad, S., … & Ali, A. (2020). Intelligent machine learning approach for effective recognition of diabetes in E-healthcare using clinical data. *Sensors*, 20(9), 2649.

Hofmann, M., & Klinkenberg, R. (Eds.). (2016). *Rapid Miner: Data mining use cases and business analytics applications*. CRC Press.

Indoria, P., & Rathore, Y. K. (2018). A survey: Detection and prediction of diabetes using machine learning techniques. *International Journal of Engineering Research & Technology (IJERT)*, 7(3), 287–291.

Islam, M. A., & Jahan, N. (2017). Prediction of onset diabetes using machine learning techniques. *International Journal of Computer Applications, 180*(5), 7–11.

Jungermann, F. (2009, February). Information extraction with RapidMiner. In *Proceedings of the GSCL Symposium Sprachtechnologie und eHumanities* (pp. 50–61).

Kad, M., & Kumbhar, C. (2019). Analysis of Summer Season and Detection of Diseases Due To Summer Season. *Journal of Emerging Technologies and Innovative Research*, Volume 6, Issue 5, 1–7.

Kopitar, L., Kocbek, P., Cilar, L., Sheikh, A., & Stiglic, G. (2020). Early detection of type 2 diabetes mellitus using machine learning-based prediction models. *Scientific Reports, 10*(1), 1–12.

Kotas, C., Naughton, T., & Imam, N. (2018, January). A comparison of Amazon Web Services and Microsoft Azure Cloud platforms for high performance computing. In *2018 IEEE International Conference on Consumer Electronics (ICCE)* (pp. 1–4). IEEE.

Kotu, V., & Deshpande, B. (2014). *Predictive analytics and data mining: Concepts and practice with RapidMiner*. Morgan Kaufmann.

Kumar Dewangan, A., & Agrawal, P. (2015). Classification of diabetes mellitus using machine learning techniques. *International Journal of Engineering and Applied Sciences, 2*(5), 257905.

Kumari, V. A., & Chitra, R. (2013). Classification of diabetes disease using support vector machine. *International Journal of Engineering Research and Applications, 3*(2), 1797–1801.

Mujumdar, A., & Vaidehi, V. (2019). Diabetes prediction using machine learning algorithms. *Procedia Computer Science, 165*, 292–299.

Mund, S. (2015). *Microsoft Azure machine learning*. Packt Publishing Ltd.

Priya, R., & Aruna, P. (2013). Diagnosis of diabetic retinopathy using machine learning techniques. *ICTACT Journal on Soft Computing, 3*(4), 563–575.

Reddy, D. J., Mounika, B., Sindhu, S., Reddy, T. P., Reddy, N. S., Sri, G. J., ... & Kora, P. (2020). Predictive machine learning model for early detection and analysis of diabetes. *Materials Today: Proceedings*.

Ristoski, P., Bizer, C., & Paulheim, H. (2015). Mining the web of linked data with RapidMiner. *Journal of Web Semantics, 35*, 142–151.

Roychowdhury, S., Koozekanani, D. D., & Parhi, K. K. (2013). DREAM: Diabetic retinopathy analysis using machine learning. *IEEE Journal of Biomedical and Health Informatics, 18*(5), 1717–1728.

Rubaiat, S. Y., Rahman, M. M., & Hasan, M. K. (2018, December). Important feature selection & accuracy comparisons of different machine learning models for early diabetes detection. In *2018 International Conference on Innovation in Engineering and Technology (ICIET)* (pp. 1–6). IEEE.

Samant, P., & Agarwal, R. (2018). Machine learning techniques for medical diagnosis of diabetes using iris images. *Computer Methods and Programs in Biomedicine, 157*, 121–128.

Sharma, N., & Singh, A. (2018, July). Diabetes detection and prediction using machine learning/ IoT: A survey. In *International Conference on Advanced Informatics for Computing Research* (pp. 471–479). Springer, Singapore.

Sisodia, D., & Sisodia, D. S. (2018). Prediction of diabetes using classification algorithms. *Procedia Computer Science, 132*, 1578–1585.

Swapna, G., Vinayakumar, R., & Soman, K. P. (2018). Diabetes detection using deep learning algorithms. *ICT Express, 4*(4), 243–246.

Verma, T., Renu, R., & Gaur, D. (2014). Tokenization and filtering process in RapidMiner. *International Journal of Applied Information Systems, 7*(2), 16–18.

Wei, S., Zhao, X., & Miao, C. (2018, February). A comprehensive exploration to the machine learning techniques for diabetes identification. In *2018 IEEE 4th World Forum on Internet of Things (WF-IoT)* (pp. 291–295). IEEE.

Wilder, B. (2012). *Cloud architecture patterns: Using Microsoft Azure*. O'Reilly Media, Inc.

Yahyaoui, A., Jamil, A., Rasheed, J., & Yesiltepe, M. (2019, November). A decision support system for diabetes prediction using machine learning and deep learning techniques. In *2019 1st International Informatics and Software Engineering Conference (UBMYK)* (pp. 1–4). IEEE.

Zou, Q., Qu, K., Luo, Y., Yin, D., Ju, Y., & Tang, H. (2018). Predicting diabetes mellitus with machine learning techniques. *Frontiers in Genetics, 9*, 515.

5

Advanced Agricultural Systems: Identification, Crop Yields and Recommendations Using Image-Processing Techniques and Machine-Learning Algorithms

Avali Banerjee and Shobhandeb Paul

Guru Nanak Institute of Technology (GNIT), Affiliated under Maulana Abul Kalam Azad University of Technology (MAKAUT), West Bengal, India

Soumi Bhattacharya

Narula Institute of Technology (NiT), Affiliated under Maulana Abul Kalam Azad University of Technology (MAKAUT), West Bengal, India

CONTENTS

DOI: 10.1201/9781003240310-5

5.1 Introduction

The world has seen many changes from the ancient and medieval to the modern period, whether in agriculture, healthcare, or science. This chapter focuses on the agriculture sector, which has adopted many modern techniques to increase crop yield, and hence also productivity and profitability. An increase in crop production serves the people of the nation and prevents food shortages. In this scenario, the price of the crops remains stable, and farmers are able to earn their living. Since the Industrial Revolution, traditional techniques have been replaced with modern techniques, including the use of fertilizers and heavy farming equipment. While advanced technologies increase the complexity of farming processes, these remain user friendly and are becoming faster and more reliable, leading to a sustainable and profitable system that takes account of environmental concerns while feeding populations.

5.2 Literature Survey

Veronica Saiz-Rubio et al. [1] discussed advanced farm management systems that include robotics and artificial intelligence. Farhat Abbas et al. [2] predicted the yield of a potato crop by applying four different machine-learning algorithms: linear regression, elastic net, k-nearest neighbour (k-NN), and support vector regression. Data were collected through proximal sensing. They observed that all ML algorithms worked well, but the performance of k-NN is sub-standard. Bhanumathi [3] developed a model with machine-learning algorithms that predict crop failure and improve revenues by increasing the crop yield according to the fertilizer ratio. which is based on basic parameters like atmosphere and soil.

Anna Chlingaryan et al. [4] discussed the accurate prediction of crop yield and nitrogen status estimation using machine-learning techniques. They combined sensing and ML techniques, leading to precision agriculture which provides better crop management solutions. Chaithra M. Rao et al. [5] observed that Random Forest Regression accurately predicts crop yields according to parameters such as weather and location. They concluded that a simple recurrent neural network performs better for rainfall prediction and LSTM is appropriate for temperature prediction. Tanha Talaviya et al. [6]review methods of enhancing quality and productivity in agriculture by using AI applications such as irrigation, weeding, and spraying by sensors embedded in robots and drones.

Ashok Tatapudi et al. [7] proposed an IoT-based remote sensing system to enhance crop productivity and determine which crop to produce effectively. Machine-learning algorithms such as linear regression, decision trees, random forest, and GD Boost were applied to data obtained from sensors, including pH, moisture, rainfall, temperature, and humidity. Shrinivas et al. [8] developed a robotic cultivator vehicle that detects obstacles. Arpit Mittal et al. [9], focusing on solutions to the drawbacks of less advanced technologies, monitored important parameters—temperature, humidity, and water level—for crop growth and maintained a database to ensure crop health. Prachi Singh et al. [10] discussed hyperspectral remote sensing which provides exact information about geographical features for sustainable agriculture.

Mohamad M. Awad [11] developed a model based on trust-region methods for a non-linear minimization algorithm to improve accuracy of crop yield estimates. Alessandro Matese et al. [12] described a multi-sensor unmanned aerial vehicle (UAV) system required for remote sensing. Corentin Leroux et al. [13] evaluated the performance of the GeoFIS open-source software which is required to ease the transfer from spatial data to spatial information.

Jash Doshi et al. [14] proposed a product to help farmers access live field data to enhance crop yields and save resources. AkshayAtole et al. [15] designed an IoT-based smart farming system, wireless sensor networks, and cloud computing to improve farming methods.

5.3 Proposed Machine-Learning System

The dataset relating to suitable crop yields and outcomes is prepared by collecting information from different sources. It is then trained using various classification algorithms, the results are compared, and the best model chosen. The machine knows the parameters responsible for better yields of a particular crop, and the conditions necessary to continue production of these yields.

The next step is to identify the type of crops to which the machine-learning model is to be applied. Image-processing techniques are used to achieve the most exact identification of the crops possible, so that the appropriate machine-learning model can be applied.

If both methods are combined, minute details of the crops will be available to the famers, which is manually not possible. This in turn helps current and future farmers to take the necessary steps to avoid losses.

Feedback is obtained from farmers to create a secure data bank that can be used for future results and to optimize the results of the existing model over time. For the future, blockchain can be used to enable secure and efficient user data transfer across platforms and systems. This technology can also be used for record keeping.

5.4 Dataset

The dataset is collected from the Kaggle community. An extract from the dataset is shown in Figure 5.1.

For the purpose of our experiment, we have collected data on major crops: wheat, rice, maize, sugarcane, and jute. To predict crops and determine optimal soil and climatic conditions, many other crops are taken into account (see Figure 5.2).

The appropriate growing conditions for specific crops can be found in the dataset. Here only one crop is listed, but many other major crops are available in the dataset to train the model. The remainder of this chapter will discuss these two datasets and analyze the results obtained from the respective models. A deep neural network model is used for training the images and a major classification algorithm is used for training the CSV file. Although the dataset looks clean, some data pre-processing still needs to be done to obtain a crisp dataset.

5.4.1 Data Pre-Processing

Data pre-processing helps us not only to understand the dataset but also to get to know its nature and the patterns within it. This in turn helps the user to develop, or rather plan, a problem set so that a better version of the information can be extracted from the dataset.

During pre-processing it is very important to check for imbalances in the dataset, such as missing values with respect to a particular attribute, which could present a problem while training the dataset. A crisp image dataset was prepared using the one hot encoding method (Figure 5.3). Many optimization techniques are available to improve data pre-processing.

```
[→  Text(0.5, 1.0, 'wheat')
```

FIGURE 5.1
Crop images from the proposed dataset.

	N	P	K	temperature	humidity	ph	rainfall	label
0	90	42	43	20.879744	82.002744	6.502985	202.935536	rice
1	85	58	41	21.770462	80.319644	7.038096	226.655537	rice
2	60	55	44	23.004459	82.320763	7.840207	263.964248	rice
3	74	35	40	26.491096	80.158363	6.980401	242.864034	rice
4	78	42	42	20.130175	81.604873	7.628473	262.717340	rice

FIGURE 5.2
Crop growth conditions dataset.

5.4.2 Train-Test Split

The next step, the train-test split, is performed by deciding labels and features of the dataset. Labels refers to the target, or y variable, shown along the y-axis; features refers to the independent variables determining the target, shown along the x-axis. The dataset is split into train and test datasets. A standard ratio (say 75%) becomes the train dataset and the rest (say 25%) is the test dataset. The training dataset is then attached to a machine-learning algorithm to obtain the best model, which is then predicted with respect to the test dataset, which measures accuracy.

Let us now visualize the model summary obtained after fitting the data into our neural networks model.

```
N                    int64
P                    int64
K                    int64
temperature        float64
humidity           float64
ph                 float64
rainfall           float64
label               object
dtype: object
```

FIGURE 5.3
Data types from the growth conditions dataset.

Model: "sequential"

Layer (type)	Output Shape	Param #
vgg19 (Functional)	(None, 512)	20024384
dense (Dense)	(None, 1000)	513000
dense_1 (Dense)	(None, 1000)	1001000
dense_2 (Dense)	(None, 1000)	1001000
dense_3 (Dense)	(None, 5)	5005

```
Total params: 22,544,389
Trainable params: 2,520,005
Non-trainable params: 20,024,384
```

FIGURE 5.4
Model summary for the VGG-19 model.

5.4.3 Creating the Classifier Model Using VGG-19

The model is initialized with VGG-19, imported from the Keraslibrary, using suitable parameters to get the best out of the model training. From the model summary (Figure 5.4), we can conclude the following key areas of the dataset:

Total Parameters available: 22,544,389

Trainable Parameters: 5,520,005

Non-trainable Parameters: 200,254,384

```
score = vggmodel.evaluate(dx_test,y_test)
print("accuracy: ", score[1])

6/6 [==============================] - 103s 16s/step - loss: 0.3616 - accuracy: 0.9255
accuracy:  0.9254658222198486
```

FIGURE 5.5
VGG-19 model evaluation and accuracy measure.

5.4.4 Evaluating the Model

A sequential model is a pile of layers (preferably plain) where every layer has distinct input and output variables.

The sequential model is not suitable under any of the following conditions:

- The model is subjected to multiple input and output.
- The layers have several inputs/outputs.
- Layer sharing is essential
- When a non-linear hierarchy is required (which can be a residual connection or the multi-branch model)

The next step is evaluation of the model for accuracy after training using the neural network model.

Figure 5.5 demonstrates that the VGG model used as the neural networks performed quite well, giving 92.54% accuracy.

The term "epoch" is used in deep-learning and machine-learning techniques to indicate the number of passes for an entire training dataset of the proposed algorithm that have been completed. For instance, when the batch size comprises the whole training dataset, then the number of epochs is calculated by the number of iterations.

After training the classification model just created, and keeping epochs = 50, the model accuracy is as shown in Figure 5.6.

Once the model performance has been established, it is time to evaluate how well our proposed model is able to identify the crop images that are being fed into the system.

From the images in Figure 5.7, we can conclude that the model is well trained according to our requirements, and can easily distinguish between crops with a good accuracy level. Using rice as our experimental sample, test prediction results are shown in Figure 5.8. The accuracy is 99.87%, meaning that the VGG-19 neural networks model has satisfied our requirement for the purpose of this chapter.

```
Epoch 44/50
29/29 [==============================] - 404s 14s/step - loss: 0.0039 - accuracy: 1.0000 - val_loss: 0.2374 - val_accuracy: 0.9430
Epoch 45/50
29/29 [==============================] - 404s 14s/step - loss: 0.0028 - accuracy: 1.0000 - val_loss: 0.2544 - val_accuracy: 0.9326
Epoch 46/50
29/29 [==============================] - 415s 14s/step - loss: 0.0028 - accuracy: 1.0000 - val_loss: 0.2736 - val_accuracy: 0.9378
Epoch 47/50
29/29 [==============================] - 408s 14s/step - loss: 0.0017 - accuracy: 1.0000 - val_loss: 0.2580 - val_accuracy: 0.9326
Epoch 48/50
29/29 [==============================] - 404s 14s/step - loss: 0.0040 - accuracy: 1.0000 - val_loss: 0.2379 - val_accuracy: 0.9326
Epoch 49/50
29/29 [==============================] - 408s 14s/step - loss: 0.0018 - accuracy: 1.0000 - val_loss: 0.2500 - val_accuracy: 0.9326
Epoch 50/50
29/29 [==============================] - 405s 14s/step - loss: 0.0023 - accuracy: 1.0000 - val_loss: 0.2729 - val_accuracy: 0.9326
```

FIGURE 5.6
Epochs training and its accuracy measure.

FIGURE 5.7
Predicted crop images after model training and evaluation.

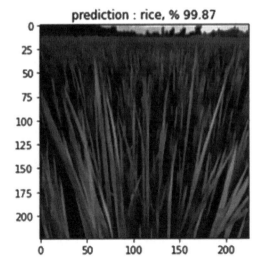

FIGURE 5.8
Rice crop image prediction with 99.87% after model training and evaluation.

5.5 Confusion Matrix

The confusion matrix is mainly used to assess the proficiency or rather the achievement of the trained classification models for the proposed test data. It indicates whether true values for the test data are known to the model. It is also referred to as an error matrix, as it gives errors within the model performance in the form of a matrix. The confusion matrix is easy to visualize and simple to understand. Its salient features are:

- Consider a case where two prediction classes of given classifiers are present, and the given matrix consists of 2×2 table, for 3 classes, 3× 3 table, and so on.
- The matrix is further split in two dimensions: the predicted values and actual values accompanying the total number of predictions.
- Predicted values are those that are predicted by the model and actual values are the true values for the given instances.
- Various calculations can be performed with the help of this matrix, e.g., model accuracy. These are described below.

Classification Accuracy: This is one of the significant specifications determining the accuracy of the given classification problem statements. It is defined as how frequently the model is capable of performing the prediction and achieving the correct output. The same method can be calculated as the ratio of the number of correct predictions made by the classifier model to the total/overall number of predictions done by the classifiers. The formula for this is:

$$\text{Accuracy} = \frac{\text{True Positive} + \text{True Negative}}{\text{True Postive} + \text{False Positive} + \text{False Negative} + \text{True Negative}} \quad (5.1)$$

Misclassification rate: This is termed the error rate, and defines how frequently the proposed model gives incorrect predictions. The evaluation of the error rate can be predicted, or rather calculated, as the number of incorrect instances to the total number of instances made by the proposed classifier. The formula is:

$$\text{Error Rate} = \frac{\text{False Positive} + \text{False Negative}}{\text{True Postive} + \text{False Positive} + \text{False Negative} + \text{True Negative}} \quad (5.2)$$

Precision: This is defined as the number of correct output/outputs made available by the model or out of total positive classes that are correctly predicted by the model, out of how many of them were actually true. The formula is:

$$\text{Precision} = \frac{\text{True Positive}}{\text{True Postive} + \text{False Positive}} \quad (5.3)$$

Recall: This is defined as the output of total positive classes, which indicates how our given proposed classifier model is predicted correctly. The value of the recall must be as high as possible. The formula is:

$$\text{Recall} = \frac{\text{True Positive}}{\text{True Postive} + \text{False Negative}} \quad (5.4)$$

F-measure: It is quite complex to compare any two models that have high recall and low precision, or vice versa. F-score can be used as a solution, as it helps us to calculate and assess the precision and recall at the same time. The F-score is at its maximum when the recall is equivalent to the precision. The formula is:

$$F-measure = \frac{2*Recall*Precision}{Recall+Precision} \qquad (5.5)$$

5.5.1 VGG-19 Model Confusion Matrix

True positive (TP), false positive (FP), true negative (TN), and false negative (FN) measures indicate how well the algorithm has trained for the proposed model and how well the train and the test data match, and these can be depicted graphically. The confusion matrix for our proposed VGG-19 model is shown in Figure 5.9.

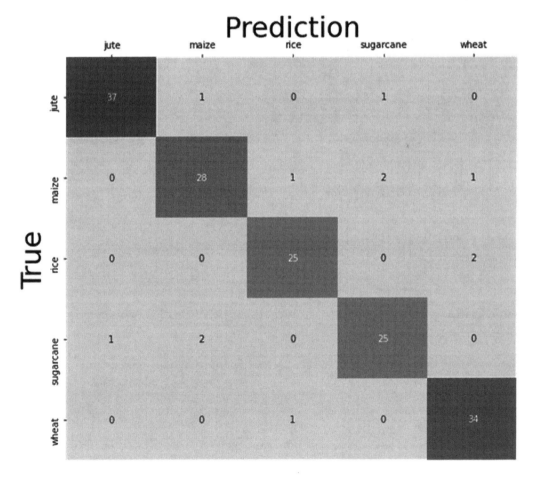

FIGURE 5.9
Confusion matrix for the VGG-19 model.

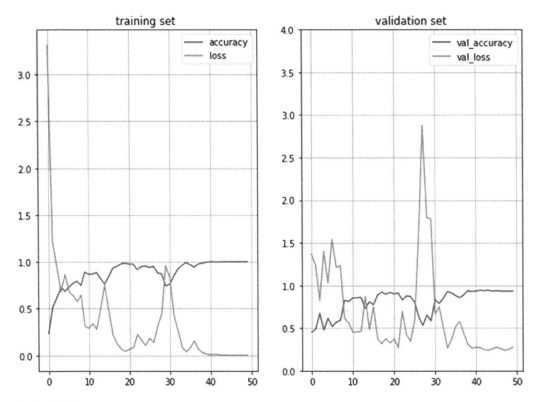

FIGURE 5.10
Accuracy and loss of training set and validation set respectively.

5.5.2 Train-Test and Validation Loss

After the model has trained itself, it is essential for performance evaluation to get an idea of the loss and the accuracy This tells us how well the model has performed for the given dataset, with a measure of loss and accuracy, which can later be used as the primary measure to compare the models.

Figure 5.10 shows how the VGG-19 model performed in order to train itself to meet our requirements.

5.6 Classification Algorithm

Classification can be described as the process of how to recognize, understand, and group objects and ideas into significant categories also known as "sub-populations". It is a form of "pattern recognition". By applying these pre-arranged training datasets, machine-learning models use various algorithms for the classification of future datasets into their respective categories, enabling patterns to be observed.

Some important classification algorithms are:

- Naive Bayes
- Random forest classifier

- Support vector machines
- XG-boost
- Decision tree classifier

We will now examine another dataset from the same source, the Kaggle community. This dataset contains all the necessary details about the key conditions required to grow a particular crop (rice in this example).

We have used various algorithms to prepare our required model. Our main concern was to train the dataset in such a way that the model could give an accuracy of at least 90%. As the dataset is of the classification type, henceforth the type of algorithms chosen to train the model was based upon classification. The following sections describe the various algorithms used and their mechanisms.

5.6.1 XG-boost

XG-boost is one of the most popular and widely used classification algorithms for machine-learning problem statements. The problem may be classified as a classification or a regression problem. It is known for its overall good performance compared to other

```
XGBoost's Accuracy is:  0.9931818181818182
                 precision    recall  f1-score   support

       apple        1.00      1.00      1.00        13
      banana        1.00      1.00      1.00        17
   blackgram        1.00      1.00      1.00        16
    chickpea        1.00      1.00      1.00        21
     coconut        1.00      1.00      1.00        21
      coffee        0.96      1.00      0.98        22
      cotton        1.00      1.00      1.00        20
      grapes        1.00      1.00      1.00        18
        jute        1.00      0.93      0.96        28
  kidneybeans        1.00      1.00      1.00        14
      lentil        0.96      1.00      0.98        23
       maize        1.00      1.00      1.00        21
       mango        1.00      1.00      1.00        26
    mothbeans        1.00      0.95      0.97        19
    mungbean        1.00      1.00      1.00        24
   muskmelon        1.00      1.00      1.00        23
      orange        1.00      1.00      1.00        29
      papaya        1.00      1.00      1.00        19
   pigeonpeas       1.00      1.00      1.00        18
  pomegranate       1.00      1.00      1.00        17
        rice        0.94      1.00      0.97        16
   watermelon       1.00      1.00      1.00        15

    accuracy                            0.99       440
   macro avg        0.99      0.99      0.99       440
weighted avg        0.99      0.99      0.99       440
```

FIGURE 5.11
Classification report for XG-boost classifier.

```
DecisionTrees's Accuracy is:  90.0
                precision    recall  f1-score   support

        apple       1.00      1.00      1.00        13
       banana       1.00      1.00      1.00        17
    blackgram       0.59      1.00      0.74        16
     chickpea       1.00      1.00      1.00        21
      coconut       0.91      1.00      0.95        21
       coffee       1.00      1.00      1.00        22
       cotton       1.00      1.00      1.00        20
       grapes       1.00      1.00      1.00        18
         jute       0.74      0.93      0.83        28
   kidneybeans       0.00      0.00      0.00        14
       lentil       0.68      1.00      0.81        23
        maize       1.00      1.00      1.00        21
        mango       1.00      1.00      1.00        26
     mothbeans       0.00      0.00      0.00        19
     mungbean       1.00      1.00      1.00        24
    muskmelon       1.00      1.00      1.00        23
       orange       1.00      1.00      1.00        29
       papaya       1.00      0.84      0.91        19
    pigeonpeas       0.62      1.00      0.77        18
  pomegranate       1.00      1.00      1.00        17
         rice       1.00      0.62      0.77        16
    watermelon       1.00      1.00      1.00        15

     accuracy                           0.90       440
    macro avg       0.84      0.88      0.85       440
 weighted avg       0.86      0.90      0.87       440
```

FIGURE 5.12
Classification report for decision tree classifier.

machine-learning algorithms. It is also known as "an enhanced gradient boosting library", making use of a specific framework, a gradient boosting framework. It performs well in neural networks when applied to prediction problems which involve unstructured datasets like images and texts. The XG-boost algorithm is the most commonly used for supervised learning problem statements in machine learning. The classification report obtained from the XG- boost model is shown in Figure 5.11, below:

5.6.2 Decision Tree

The decision tree algorithm is one of the most widely used classification algorithms that fall into the category of supervised learning. Decision trees are applied to predict both regression and classification problem statements. They are usually depicted in the form of a tree in two parts or entities, which are known as the decision nodes and leaves. Every leaf node corresponds to a class label and the internal nodes of the tree correspond to the attributes. The leaves depict the decisions or final outcomes, and the nodes form a decision node where the data is split. We can represent any Boolean function on specific attributes

```
RF's Accuracy is:  0.990909090909091
                precision    recall  f1-score   support

       apple       1.00      1.00      1.00        13
      banana       1.00      1.00      1.00        17
   blackgram       0.94      1.00      0.97        16
    chickpea       1.00      1.00      1.00        21
     coconut       1.00      1.00      1.00        21
      coffee       1.00      1.00      1.00        22
      cotton       1.00      1.00      1.00        20
      grapes       1.00      1.00      1.00        18
        jute       0.90      1.00      0.95        28
  kidneybeans      1.00      1.00      1.00        14
      lentil       1.00      1.00      1.00        23
       maize       1.00      1.00      1.00        21
       mango       1.00      1.00      1.00        26
    mothbeans      1.00      0.95      0.97        19
    mungbean       1.00      1.00      1.00        24
   muskmelon       1.00      1.00      1.00        23
      orange       1.00      1.00      1.00        29
      papaya       1.00      1.00      1.00        19
   pigeonpeas      1.00      1.00      1.00        18
  pomegranate      1.00      1.00      1.00        17
        rice       1.00      0.81      0.90        16
  watermelon       1.00      1.00      1.00        15

    accuracy                           0.99       440
   macro avg       0.99      0.99      0.99       440
weighted avg       0.99      0.99      0.99       440
```

FIGURE 5.13
Classification report for random forest classifier.

using the decision tree. The two main types of decision trees are: classification trees (yes/no type), where the decision variable is categorical; and regression trees (mainly for continuous data types).

Here the outcome or the decision variable is considered continuous. The classification report of a decision tree classifier model is shown in Figure 5.12.

5.6.3 Random Forest Classifier

Random forest is also one of the most widely used supervised machine-learning algorithms. It can be applied to predict both classification and regression problems. Random Forest provides various random features which in turn helps the model to limit the error due variance and bias.

The most notable feature of the random forest algorithm, which can handle a dataset of both continuous variables, as in the case of regression, and categorical variables as in the case of classification, is that it mainly gives better results for classification problems. The classification report of the random forest classifier is shown in Figure 5.13.

```
Naive Bayes's Accuracy is:  0.990909090909091
                 precision    recall  f1-score   support

       apple       1.00      1.00      1.00        13
      banana       1.00      1.00      1.00        17
    blackgram       1.00      1.00      1.00        16
     chickpea       1.00      1.00      1.00        21
      coconut       1.00      1.00      1.00        21
       coffee       1.00      1.00      1.00        22
       cotton       1.00      1.00      1.00        20
       grapes       1.00      1.00      1.00        18
         jute       0.88      1.00      0.93        28
   kidneybeans       1.00      1.00      1.00        14
       lentil       1.00      1.00      1.00        23
        maize       1.00      1.00      1.00        21
        mango       1.00      1.00      1.00        26
     mothbeans       1.00      1.00      1.00        19
     mungbean       1.00      1.00      1.00        24
     muskmelon       1.00      1.00      1.00        23
       orange       1.00      1.00      1.00        29
       papaya       1.00      1.00      1.00        19
    pigeonpeas       1.00      1.00      1.00        18
  pomegranate       1.00      1.00      1.00        17
         rice       1.00      0.75      0.86        16
    watermelon       1.00      1.00      1.00        15

     accuracy                          0.99       440
    macro avg       0.99      0.99      0.99       440
 weighted avg       0.99      0.99      0.99       440
```

FIGURE 5.14
Classification report for naive Bayes classifier.

5.6.4 Naive Bayes Classifier

Naive Bayes is based on a famous theorem proposed by Thomas Bayes and widely applied to solve classification problem statements. The application of this algorithm can also be seen in text classification which includes a high-dimensional training dataset.

This is one of the simplest and most effective classification algorithms and is capable of making quick predictions. This is a very good example of a probabilistic classifier that functions on the basis of probability of an object. Popular examples of naive Bayes algorithms are spam detection and sentiment analysis. The classification report of the naive Bayes classifier is shown in Figure 5.14.

5.6.5 Support Vector Machine

For classification and regression problem statements, SVM proves one of the most widely applied supervised machine-learning algorithms. The SVM algorithm aims to solve the problem by representing the best-fit line, which can be divided into n-dimensional space into classes. This feature makes it easy for us to represent the new data tensor in the right form for future applications. This gives rise to an important term, "hyperplane", based on a best-fit line or decision boundary.

```
SVM's Accuracy is: 0.9795454545454545
                 precision    recall  f1-score   support

         apple      1.00      1.00      1.00        13
        banana      1.00      1.00      1.00        17
     blackgram      1.00      1.00      1.00        16
      chickpea      1.00      1.00      1.00        21
       coconut      1.00      1.00      1.00        21
        coffee      1.00      0.95      0.98        22
        cotton      0.95      1.00      0.98        20
        grapes      1.00      1.00      1.00        18
          jute      0.83      0.89      0.86        28
    kidneybeans      1.00      1.00      1.00        14
        lentil      1.00      1.00      1.00        23
         maize      1.00      0.95      0.98        21
         mango      1.00      1.00      1.00        26
      mothbeans      1.00      1.00      1.00        19
      mungbean      1.00      1.00      1.00        24
     muskmelon      1.00      1.00      1.00        23
        orange      1.00      1.00      1.00        29
        papaya      1.00      1.00      1.00        19
     pigeonpeas      1.00      1.00      1.00        18
   pomegranate      1.00      1.00      1.00        17
          rice      0.80      0.75      0.77        16
    watermelon      1.00      1.00      1.00        15

      accuracy                          0.98       440
     macro avg      0.98      0.98      0.98       440
  weighted avg      0.98      0.98      0.98       440
```

FIGURE 5.15
Classification report for support vector machine.

SVM functions by choosing the vectors that represent the hyper plane. These vectors are known as the support vectors, hence the algorithm's name of support vector machine. Take the following example. Suppose there is a strange cat which possesses some features of dogs. A model that is capable of distinguishing between both entities can be developed using this classification algorithm. First our model must be trained with a good number of images, preferably high-resolution images of cats and dogs, so that the machine is exposed to different characteristics of both entities. Next we need to make a comparison with the test data. The algorithm now takes a decision by creating a best-fit line, which selects the vectors known as support vectors. Finally, depending on the type of support vectors, the entities are classified as a cat or dog. The classification report of SVM is shown in Figure 5.15.

5.7 Conclusion

The proposed system can be considered an advanced agricultural system, as many advanced techniques are involved, from identification and detection of the crops using image processing to the decision of an appropriate and efficient machine-learning model and recommendation for the farmers regarding the crops. The whole constitutes a complete system to help the farmers to modernize. The application of this proposed model should provide improved yields for a more sustainable farming community, which is improving every day.

References

1. Veronica Saiz-Rubio, Francisco Rovira Mas, 'From Smart Farming towards Agriculture 5.0: A Review on Crop Data Management'. *Agronomy, MDPI Journal*, 10, p. 207, 2020. doi: 10.3390/agronomy10020207
2. Farhat Abbas, Hassan Afzaal, Aitazaz A. Farooque, Skylar Tang, 'Crop Yield Prediction Through Proximal Sensing and Machine Learning Algorithms'. *Agronomy, MDPI Journal*, 10, p. 1046, 2020. doi: 10.3390/agronomy10071046
3. S. Bhanumathi, M. Vineeth, N. Rohit, 'Crop Yield Prediction and Efficient Use of Fertilizers'. *International Conference on Communication and Signal Processing*, April 4-6, India. 2019. pp. 769–773. doi: 10.1109/ICCSP.2019.8698087
4. Anna Chlingaryan, Salah Sukkarieh, Brett Whelan, 'Machine Learning Approaches for Crop Yield Prediction and Nitrogen Status Estimation in Precision Agriculture: A Review'. *Computers and Electronics in Agriculture*,151, pp. 61–69. Elsevier. 2018. doi: 10.1016/j.compag.2018.05.012
5. Chaithra M. Rao, C. K. Sanjay Kumar, 'Crop Yield Prediction and Efficient Use of Fertilizers'. *International Research Journal of Engineering and Technology (IRJET)*, 07(07), July2020. pp. 2372–2377.
6. Tanha Talaviya, Dhara Shah, Nivedita Patel, Hiteshri Yagnik, Manan Shah, 'Implementation of Artificial Intelligence in Agriculture for Optimisation of Irrigation and Application of Pesticides and Herbicides'. *Artificial Intelligence in Agriculture*, 4, pp. 58–73, 2020. doi: 10.1016/j.aiia.2020.04.002
7. P. Ashok Tatapudi, Suresh Varma, 'Prediction of Crops Based on Environmental Factors Using IoT & Machine Learning Algorithms'. *International Journal of Innovative Technology and Exploring Engineering (IJITEE)* ISSN: 2278-3075, 9(1), November 2019. pp. 5395–5401.
8. Shrinivas R. Zanwar, R. D. Kokate, 'Advanced Agriculture System'. *International Journal of Robotics and Automation (IJRA)*, 1(2), June 2012, pp.107–112, 2012. https://doi.org/10.11591/ijra.v1i2.382
9. Arpit Mittal, Niladri Nandan Sarma, A. Sriram, Trisha Roy, Shriya Adhikari, 'Advanced Agriculture System Using GSM Technology'. *International Conference on Communication and Signal Processing (ICCSP)*, pp. 285–289, 3–5 April 2018. doi: 10.1109/ICCSP.2018.8524538.
10. Prachi Singh, Prem Chandra Pandey, George P. Petropoulos, Andrew Pavlides, Prashant K. Srivastava, Nikos Koutsias, Khidir Abdala Kwal Deng, Yangson Bao, 'Hyperspectral remote sensing in precision agriculture: Present status, challenges, and future trends, hyperspectral remote sensing'. *Hyperspectral Remote Sensing Theory and Applications*, Earth Observation, pp. 121–146, 2020. doi: 10.1016/B978-0-08-102894-0.00009-7
11. Mohamad M. Awad, 'Toward Precision in Crop Yield Estimation Using Remote Sensing and Optimization Techniques'. *Agriculture, MDPI Journal*, 9(3), p. 54, 2018. doi: 10.3390/agriculture9030054
12. Alessandro Matese, Salvatore Filippo DiGennaro, 'Practical Applications of a Multi sensor UAV Platform Based on Multispectral, Thermal and RGB High Resolution Images in Precision Viticulture'. *Agriculture, MDPI Journal*, 8(7), p. 116, 2018. doi: 10.3390/agriculture8070116
13. Corentin Leroux, Hazaël Jones, Léo Pichon, Serge Guillaume, Julien Lamour, James Taylor, Olivier Naud, Thomas Crestey, Jean-Luc Lablee, Bruno Tisseyre, 'GeoFIS: An Open Source, Decision-Support Tool for Precision Agriculture Data'. *Agriculture, MDPI Journal*, 8(6), p. 73, 2018. doi: 10.3390/agriculture8060073

14. Jash Doshi, Tirthkumar Patel, Santosh Kumar Bharti, 'Smart Farming Using IoT, a Solution for Optimally Monitoring Farming Conditions'. *The 3rd International Workshop on Recent Advances on Internet of Things: Technology and Application Approaches (IoT-T& A 2019)*, November 4–7, Coimbra, Portugal. Procedia Computer Science, 160, pp. 746–751, 2019. doi: 10.1016/j.procs.2019.11.016

15. Akshay Atole, Apurva Asmar, Amar Biradar, Nikhil Kothawade, Sambhaji Sarode, Rajendra G. Khope, 'IOT Based Smart Farming System'. *Journal of Emerging Technologies and Innovative Research*, 4(04), 2017. pp. 29–31, April 2017.

6

SP-IMLA: Stroke Prediction Using an Integrated Machine-Learning Approach

Amit Bairwa, Satpal Singh Kushwaha, Vineeta Soni, Prashant Hemrajani, and Sandeep Joshi
Manipal University, Jaipur, India

Pulkit Sharma
Ceridian, Toronto, Ontario, Canada

CONTENTS

6.1 Introduction

According to the World Health Organization (WHO), stroke is the second greatest cause of loss of life globally, accounting for about 12% of deaths [1]. With 500–900 strokes per 100,000, population, 16 million new acute strokes per year, 29 million years of life impacted by incapacity, and a 29–30-time case casualty rate varying from 18% to 36%, stroke is the second-highest cause of death and impairment in adults [2]. Stroke is predicted to grow increasingly prevalent, with 6 million deaths from stroke and heart disease estimated by 2021, up from 3 million in 1999 [3]. This will occur because of ongoing health and demographic changes, rising vascular disease risk factors, and an ageing population [4]. Developing countries account for 86% of all stroke deaths worldwide. Stroke has far-reaching social and economic consequences [5, 6]. Co-morbidities and complications are the leading causes of stroke death. Complications from a stroke may occur at any time [7].

DOI: 10.1201/9781003240310-6

The first week and the first month after the onset of stroke symptoms are critical for survival, with many fatalities occurring in the first week [8]. Death is generally caused by transtentorial herniation and haemorrhage within the first week after a stroke, bleeding within the first three days [9], and cerebral infarction between the third and sixth days [10]. The most common causes of mortality one week after a stroke are problems caused by relative immobility, such as pneumonia, sepsis, and pulmonary embolism [11].

Even though stress is linked to depression and women report stronger emotional responses to stressful life events, the link between stress and gender in relation to post-stroke depression (PSD) has yet to be investigated. PSD is linked to poor clinical and functional results, and women are more likely to be affected than men. There are also other factors to consider, such as a family history of heart disease, marital status, and whether you live in a rural or urban location, as well as employment status, such as private, government, self-employed, or unemployed. The goal of this study is to look at the numerous elements that are thought to be the most important in causing a stroke. These factors are represented statistically in Figures 6.1–6.6.

Various studies have discovered multiple variables linked to stroke mortality in their respective settings. Previous stroke, atrial fibrillations, and hypertension, for example, were the most prevalent predictors of mortality from stroke in people over 65 years old, according to Mackay [12]. Fixed dilated pupil(s), a Glasgow coma score of less than 10 on admission, swallowing problems, fever, lung infection, and no aspirin therapy were all independent risk factors for a fatal outcome in patients with haemorrhagic stroke [10]. At 7 days, 30 days, and one year following a stroke, [13] found that stroke severity, neurological deterioration during hospitalization, non-use of antithrombotic at hospital admission, and absence of assessment by a stroke team were the most consistent predictors of case mortality. Case fatality at 30 days was substantially greater in Pretoria, South Africa than in high-income countries, with 22% for ischaemic stroke and 58% for cerebral haemorrhagic stroke, and hypertension was strongly linked with stroke [14].

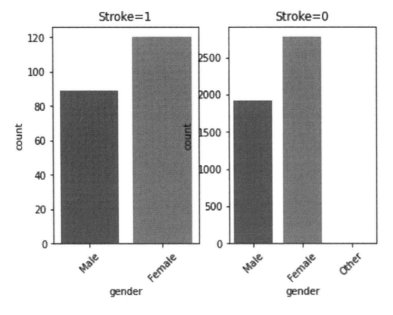

FIGURE 6.1
Gender-based observation.

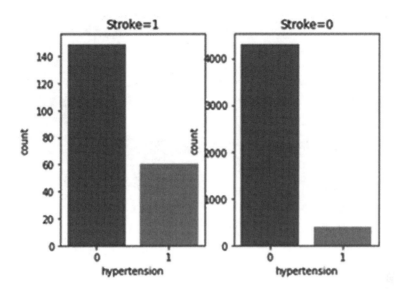

FIGURE 6.2
Observation based on hypertension.

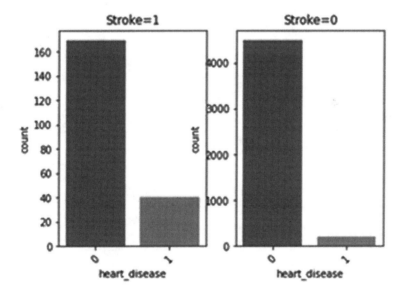

FIGURE 6.3
Observation based on heart disease.

6.1.1 Traditional Risk Factors for Stroke

Stroke may strike anybody, regardless of colour, gender, or age; nevertheless, the odds of having a stroke rise if a person has specific risk factors for stroke [15]. According to studies, understanding personal risk and how to manage it can prevent 80% of strokes. There are two types of stroke risk factors: modifiable and non-modifiable [16]. Lifestyle risk factors and medical risk factors are two types of modifiable risk factors. Alcohol, smoking, physical inactivity, and obesity are all lifestyle risk factors that may

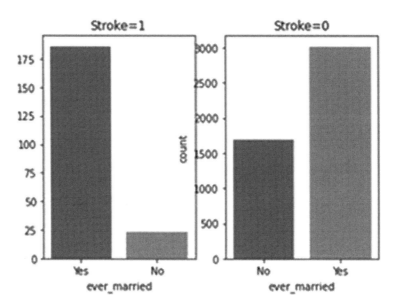

FIGURE 6.4
Observation based on marital status.

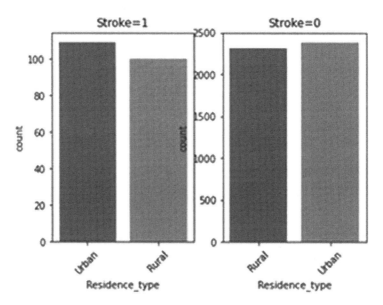

FIGURE 6.5
Observation based on residence type.

be altered [17]. Medical risk factors that can typically be addressed are diabetes mellitus, high blood pressure, atrial fibrillation, and excessive cholesterol. A primary multi-centre case-control research study (INTERSTROKE) found eleven variables linked to 90% of stroke risk, half of which are modifiable [18]. Non-modifiable risk factors, on the other hand, cannot be changed, but they can assist in identifying those who are at risk for stroke [19] (Figures 6.7–6.12).

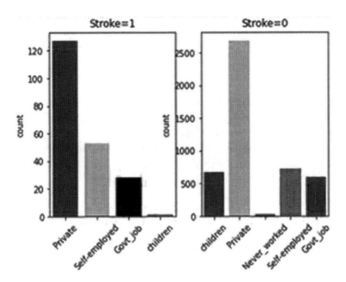

FIGURE 6.6
Observation based on job role.

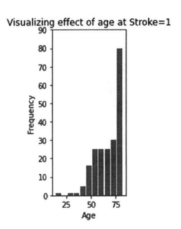

FIGURE 6.7
Observation 1 based on effect on age.

6.1.2 Stroke Prevention

More than 70% of strokes are first-time events, which makes primary stroke prevention critical [20]. Interventions should target behaviour modification, which requires information about the baseline perceptions, knowledge, and prevalence of risk factors in defined populations.

6.2 Problem Statement

Stroke is the second leading cause of death in the world, and it remains one of the serious public health issues that affects both individuals and national healthcare systems [21]. Stroke risk factors include hypertension, cardiovascular disease, diabetes, glucose

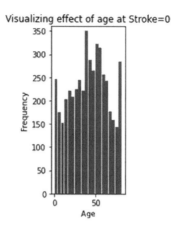

FIGURE 6.8
Observation 2 based on effect on age.

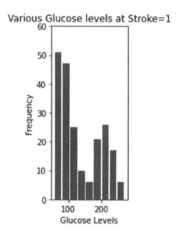

FIGURE 6.9
Observation 1 based on effect on glucose level.

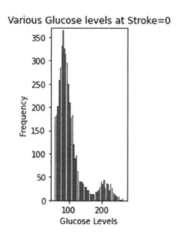

FIGURE 6.10
Observation 2 based on glucose level.

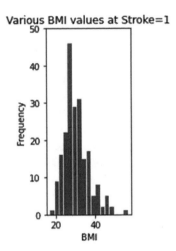

FIGURE 6.11
Observation 1 based on BMI.

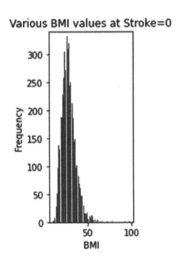

FIGURE 6.12
Observation 2 based on BMI.

dysregulation, atrial fibrillation, and lifestyle variables [17]. The goal of our research is to accurately predict stroke using machine-learning approaches based on potentially modifiable risk factors [22] and risk variables using existing huge datasets. Cheil Hong Kong planned to create an app, in collaboration with The Hong Kong Stroke Association, that would offer a tailored warning based on each user's stroke risk level as well as a lifestyle corrective message regarding stroke risk factors [23].

The main cause of stroke, a life-threatening medical condition [24], is disturbance in blood movement depriving brain muscle of oxygen and nutrients. Brain cells are damaged or die in fractions of minutes, compared to other organs [25]. Risk factors for stroke include smoking, diabetes, age, obesity, hypercholesterolemia, and hypertension.

The dataset aims to forecast if a patient is likely to undergo a stroke based on feedback factors such as sex, stage, various illnesses, and smoking level, as described in Figures 6.7–6.12. The patient's data is included in each line of data in the menu [26].

6.3 Motivation

Emerging countries are plagued by infectious and non-communicable diseases [12]. Stroke is a significant cause of early mortality and disability in nations like India, owing to changing demographics and the rising incidence of crucial modifiable risk factors [27]. Because of changing population exposure to risk factors and, most disastrously, their lack of ability to pay the high costs of stroke treatment, the weak are disproportionately affected by stroke [28]. For doctors, it is highly desirable to anticipate the likelihood of a stroke [29], enabling them to set acceptable goals with patients and families and reach standard recovery or rehabilitation decisions before treatment [20]. The purpose of this research is to learn more about strokes and develop a stroke prediction model.

6.4 Objectives of the Study

Stroke is the third-highest cause of mortality in the United States and the leading cause of significant long-term incapacity. It most commonly affects people aged 65 and over. However, it may now also affect younger populations owing to poor lifestyle. Accurate stroke prediction is crucial for early intervention and therapy. Our recommended feature selection approach generates a more extensive area under the ROC curve when coupled with support vector machines (SVMs) (AUC). This is a substantially improvement over the L1 regularized Cox feature selection method and the Cox proportional hazards model. We also looked at a margin-based censored regression approach that combines margin-centred classifier with suppressed reversion to generate a higher concordance indicator than the Cox version. Our technique is superior to existing state-of-the-art techniques in terms of concordance index and AUC metrics. We also highlight potential risks yet to be discovered using traditional methods. Our approach can predict the clinical outcome of various illnesses when risk factors are not well known and data is missing.

```
#prediction using kmeans and accuracy
kpred = KMeans_Clustering.predict(X_test)
print('Classification report:\n\n', sklearn.metrics.classification_report(y_test,kpred))
```

```
Classification report:

              precision    recall  f1-score   support

           0       0.96      0.87      0.91       929
           1       0.12      0.30      0.17        53

    accuracy                           0.84       982
   macro avg       0.54      0.58      0.54       982
weighted avg       0.91      0.84      0.87       982
```

On the Cardiovascular Health Study (CHS) dataset, five different machine-learning algorithms were employed to predict stroke in this research. Dimension reduction is accomplished by principal component analysis (PCA), feature selection is performed through a decision tree (DT) with the C4.5 method, and classification is accomplished through artificial neural network (ANN) and support vector machine (SVM). The prediction approaches presented in this work were tested on different datasets using various machine-learning algorithms, and the mixed method of PCA, DT, and ANN produced the best results.

Every year, a large portion of the world's population, about 26 million people, suffers from heart failure. Predicting heart failure at the right moment, whether from the perspective of a cardiac consultant or a surgeon, is a complicated task. Fortunately, the medical profession may benefit from different machine-learning approaches and the presence of neural networks, which can see how to use medical data efficiently. One of this study's goals was to improve the accuracy of heart failure prediction using the UCI heart disease dataset. Many standard machine-learning methods are utilized to interpret the data and forecast the risks of heart failure in a medical database.

6.5 Review of Relevant Literature

In the course of this study we reviewed research articles comparable to the work we are pursuing, learning about the authors' aims, the strategies they used to accomplish their project, and why they picked those approaches. This will help us complete our project structure by determining which techniques to employ, and how to make our project as efficient as possible.

6.6 Methodology

System architecture:

RAM	Minimum 8–16 Gb
CPU	Intel Core i5 7th-generation or higher is preferred.
Storage	Minimum 256GB or more (1TB preferred).
Operating system	Linux/Windows (preferred)
Internet	High-speed internet connectivity required.

6.7 Technology Used

Some basic knowledge of data science is needed together with the following:

- Python
- Skit-Learn
- Pandas
- Matplotlib
- Jupyter Notebook

TABLE 6.1

Pros & Cons: K-Means

PROS	CONS
Relatively simple to implement.	Choosing k manually
Scales to large datasets	Being dependent on initial values
Guarantees convergence	Scaling with number of dimensions
Can warm-start the positions of centroids	Clustering outliers
Easily adapts to new examples	Clustering data of varying sizes and density

6.8 Algorithms/Techniques

6.8.1 K-Means

This is either a distance-based algorithm or a centroid-based method. Each cluster in k-means is connected with a centroid (Table 6.1).

6.8.2 Logistic Regression

In its very basic style, logistic reversion is an arithmetic model that uses logistic function to explain a binary dependent variable, although there are many more complicated versions. Logistic relapse is used while the reliance on variable (goal) is clear-cut. There is no need to scale the features. It does not involve mounted input elements and highly values data presentation. However, all the essential features must be recognized for it to work correctly (Figure 6.13).

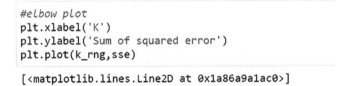

```
#elbow plot
plt.xlabel('K')
plt.ylabel('Sum of squared error')
plt.plot(k_rng,sse)
```

[<matplotlib.lines.Line2D at 0x1a86a9a1ac0>]

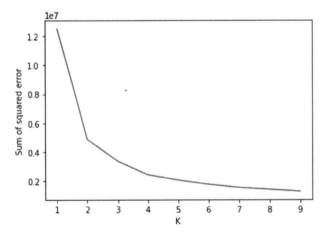

FIGURE 6.13
Finding k value: Elbow plot.

```
logreg.fit(X_train,y_train)
y_pred=logreg.predict(X_test)
count=0
for i in range(len(y_test)):
    if y_pred[i]==y_test[i]:
        count+=1
print(count*100/len(y_pred))
```

94.60285132382892

Out[55]: array([[153, 59],
                 [46, 163]])

In [56]:
```
TN=res[1][1]
TP=res[0][0]
FN=res[1][0]
FP=res[0][1]
print(TP,FP,FN,TN)
```

153 59 46 163

In [57]:
```
Sensitivity=(TP)/(TP+FN)
Specificity=(TN)/(TN+FP)
Precision=(TP)/(TP+FP)
Recall=(TP)/(TP+FN)
f1=2*(Precision*Recall)/(Precision+Recall)
```

In [58]:
```
print('Sensitivity: '+str(Sensitivity))
print('Specificity: '+str(Specificity))
print('Precision: '+str(Precision))
print('Recall: '+str(Recall))
print('f1: '+str(f1))
```

Sensitivity: 0.7688442211055276
Specificity: 0.7342342342342343
Precision: 0.7216981132075472
Recall: 0.7688442211055276
f1: 0.7445255474452555

```
Sensitivity=(TP)/(TP+FN)
Specificity=(TN)/(TN+FP)
Precision=(TP)/(TP+FP)
Recall=(TP)/(TP+FN)
Accuracy=(TP+TN)/(TP+TN+FP+FN)
f1=2*(Precision*Recall)/(Precision+Recall)

print('Sensitivity: '+str(Sensitivity))
print('Specificity: '+str(Specificity))
print('Precision: '+str(Precision))
print('Recall: '+str(Recall))
print('f1: '+str(f1))
print('Accuracy: '+str(Accuracy))
```

Sensitivity: 0.9561091340450771
Specificity: 0.11510791366906475
Precision: 0.8675995694294941
Recall: 0.9561091340450771
f1: 0.909706546275395
Accuracy: 0.8370672097759674

6.9 Conclusion and Future Work

We looked at a variety of machine-learning techniques to predict strokes in patients for this research. Before training, we ran some exploratory data analysis (primarily visual) and used one-hot encoding to transform categorical data into numeric data. The model was trained on the remaining data after the records with missing BMI values were eliminated. After properly pre-processing the data, we used the Scikit-learn default settings to apply two distinct machine-learning methods, logistic regression and k-means clustering. Based on the F1 score, we chose the best algorithm. The harmonic mean of accuracy and recall is given by F1 (a high F1 value indicates good precision and a good recall value). Cross-validation revealed that the k-means model performed best, with excellent scores across the board, notably in the recall. We may also be confident in our choice because of the solid F1 score. This model should be pretty accurate in predicting who is most likely to have a stroke and who does not require unnecessary treatment. In addition, we would use other approaches and compare the findings to discover the best match.

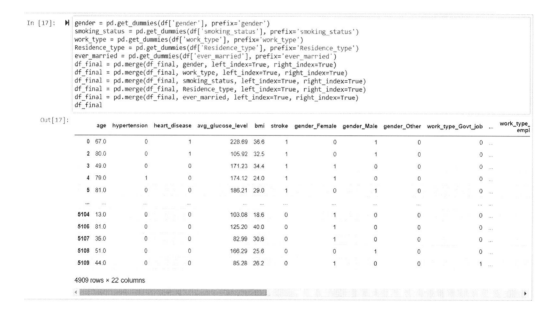

We obtained 88% accuracy with the k-means technique and 97% accuracy with the logistic regression technique. More approaches would be used, and the results would be compared to determine the best fit. We would use the random forest algorithm strategy to discover the outcome as an example of such a technique. We can use statistical models on each patient individually to monitor their health and warn them of deteriorating health conditions/possible future stroke scenarios. The goal of this project is to compare different techniques to determine the best way to predict stroke and to predict strokes more accurately and efficiently. We will also try increasing the accuracy of the result obtained using the k-means technique.

References

1. B. Singh, and H. Tawfik, A Machine Learning Approach for Predicting Weight Gain Risks in Young Adults, *Conf. Proc. 2019 10th Int. Conf. Dependable Syst. Serv. Technol. DESSERT 2019*, pp. 231–234, 2019, doi: 10.1109/DESSERT.2019.8770016

2. G. Guidi, M. C. Pettenati, P. Melillo, and E. Iadanza, A machine learning system to improve heart failure patient assistance, *IEEE J. Biomed. Heal. Informatics*, vol. 18, no. 6, pp. 1750–1756, 2014, doi: 10.1109/JBHI.2014.2337752

3. M. A. Hakim, M. Z. Hasan, M. M. Alam, M. M. Hasan, and M. M. Hasan, An Efficient Modified Bagging Method for Early Prediction of Brain Stroke, *5th Int. Conf. Comput. Commun. Chem. Mater. Electron. Eng. IC4ME2 2019*, pp. 11–12, 2019, doi: 10.1109/IC4ME247184.2019.9036700

4. M. Wang, X. Yao, and Y. Chen, An Imbalanced-Data Processing Algorithm for the Prediction of Heart Attack in Stroke Patients, *IEEE Access*, vol. 9, pp. 25394–25404, 2021, doi: 10.1109/ACCESS.2021.3057693

5. N. K. Kumar, G. S. Sindhu, D. K. Prashanthi, and A. S. Sulthana, Analysis and Prediction of Cardio Vascular Disease Using Machine Learning Classifiers, *2020 6th Int. Conf. Adv. Comput. Commun. Syst. ICACCS 2020*, no. Ml, pp. 15–21, 2020, doi: 10.1109/ICACCS48705.2020.9074183

6. G. Fang, P. Xu, and W. Liu, Automated Ischemic Stroke Subtyping Based on Machine Learning Approach, *IEEE Access*, vol. 8, pp. 118426–118432, 2020, doi: 10.1109/ACCESS.2020.3004977

7. X. Chen, C. Wei, W. Wu, L. Guo, C. Liu, and G. Lu, Based on Machine Learning Algorithm: Construction of an Early Prediction Model of Integrated Traditional Chinese and Western Medicine for Cognitive Impairment after Ischemic Stroke, *5th Int. Conf. Univers. Village, UV 2020*, 2020, doi: 10.1109/UV50937.2020.9426200

8. F. Keyrouz, L. Tauk, and E. Feghali, Chemical Structure Recognition and Prediction: A Machine Learning Technique, *2018 IEEE Conf. Comput. Intell. Bioinforma. Comput. Biol. CIBCB 2018*, pp. 1–7, 2018, doi: 10.1109/CIBCB.2018.8404964

9. Y. Oh, S. Park, and J. C. Ye, Deep Learning COVID-19 Features on CXR using Limited Training Data Sets, *IEEE Transactions on Medical Imaging*, vol. 39, no. 8, pp. 2688–2700, Aug. 2020, doi: 10.1109/TMI.2020.2993291

10. C. Y. Hung, C. H. Lin, and C. C. Lee, Improving Young Stroke Prediction by Learning with Active Data Augmenter in a Large-Scale Electronic Medical Claims Database, *Proc. Annu. Int. Conf. IEEE Eng. Med. Biol. Soc. EMBS*, vol. 2018-July, pp. 5362–5365, 2018, doi: 10.1109/EMBC.2018.8513479

11. P. Govindarajan, K. S. Ravichandran, S. Sundararajan, and S. Sreeja, Factors on the Prediction of Stroke Disease, *Int. Conf. Trends Electron. Informatics ICEI 2017*, pp. 985–989, 2017.

12. R. Ali, U. Qidwai, and S. K. Ilyas, Use of Combination of PCA and ANFIS in Infarction Volume Growth Rate Prediction in Ischemic Stroke, *2018 IEEE EMBS Conf. Biomed. Eng. Sci. IECBES 2018 - Proc.*, pp. 324–329, 2019, doi: 10.1109/IECBES.2018.8626629

13. M. Salucci, D. Marcantonio, M. Li, G. Oliveri, P. Rocca, and A. Massa, Innovative Machine Learning Techniques for Biomedical Imaging, *2019 IEEE Int. Conf. Microwaves, Antennas, Commun. Electron. Syst. COMCAS 2019*, pp. 5–7, 2019, doi: 10.1109/COMCAS44984.2019.8958253

14. N. Prentzas, A. Nicolaides, E. Kyriacou, A. Kakas, and C. Pattichis, Integrating machine learning with symbolic reasoning to build an explainable ai model for stroke prediction, *Proc. - 2019 IEEE 19th Int. Conf. Bioinforma. Bioeng. BIBE 2019*, pp. 817–821, 2019, doi: 10.1109/BIBE.2019.00152

15. M. U. Emon, M. S. Keya, T. I. Meghla, M. M. Rahman, M. S. Al Mamun, and M. S. Kaiser, Performance Analysis of Machine Learning Approaches in Stroke Prediction, *Proc. 4th Int. Conf. Electron. Commun. Aerosp. Technol. ICECA 2020*, pp. 1464–1469, 2020, doi: 10.1109/ICECA49313.2020.9297525

16. J. Cho, A. Alharin, Z. Hu, N. Fell, and M. Sartipi, Predicting Post-Stroke Hospital Discharge Disposition Using Interpretable Machine Learning Approaches, *Proc. - 2019 IEEE Int. Conf. Big Data, Big Data 2019*, pp. 4817–4822, 2019, doi: 10.1109/BigData47090.2019.9006592

17. J. Yu et al., Semantic Analysis of NIH Stroke Scale Using Machine Learning Techniques, *2019 Int. Conf. Platf. Technol. Serv. PlatCon 2019 - Proc.*, pp. 3–7, 2019, doi: 10.1109/PlatCon.2019.8668961
18. C. S. Nwosu, S. Dev, P. Bhardwaj, B. Veeravalli, and D. John, Predicting Stroke from Electronic Health Records, *Proc. Annu. Int. Conf. IEEE Eng. Med. Biol. Soc. EMBS*, pp. 5704–5707, 2019, doi: 10.1109/EMBC.2019.8857234
19. P. Kalpana, S. Shiyam Vignesh, L. M. P. Surya, and V. Vishnu Prasad, Prediction of Heart Disease Using Machine Learning, *J. Phys. Conf. Ser.*, vol. 1916, no. 1, pp. 1275–1278, 2021, doi: 10.1088/1742-6596/1916/1/012022
20. M. Subramaniyam, S. H. Hong, D. M. Kim, J. Yu, and S. J. Park, Wake-Up Stroke Prediction Through IoT and Its Possibilities, *2017 Int. Conf. Platf. Technol. Serv. PlatCon 2017 - Proc.*, 2017, doi: 10.1109/PlatCon.2017.7883738
21. K. Vijiyakumar, B. Lavanya, I. Nirmala, and S. Sofia Caroline, Random Forest Algorithm for the Prediction of Diabetes, *2019 IEEE Int. Conf. Syst. Comput. Autom. Networking, ICSCAN 2019*, 2019, doi: 10.1109/ICSCAN.2019.8878802
22. S. Capoglu, J. P. Savarraj, S. A. Sheth, H. A. Choi, and L. Giancardo, Representation Learning of 3D Brain Angiograms, an Application for Cerebral Vasospasm Prediction, *Proc. Annu. Int. Conf. IEEE Eng. Med. Biol. Soc. EMBS*, pp. 3394–3398, 2019, doi: 10.1109/EMBC.2019.8857815
23. E. Benavidez, G. B. DeMartinis, Y. Wu, and A. J. Gatesman, Application of MM-Wave Radar and Machine Learning for Post-Stroke Upper Extremity Motor Assessment, *IEEE Radar Conference (RadarConf21)*, pp. 1–6, 2021, doi: 10.1109/radarconf2147009.2021.9455191
24. X. Chen, L. Gong, L. Zheng, and Z. Zou, Soft Exoskeleton Glove for Hand Assistance Based on Human-Machine Interaction and Machine Learning, *Proc. 2020 IEEE Int. Conf. Human-Machine Syst. ICHMS 2020*, 2020, doi: 10.1109/ICHMS49158.2020.9209381
25. T. Chauhan, S. Rawat, S. Malik, and P. Singh, Supervised and Unsupervised Machine Learning Based Review on Diabetes Care, *2021 7th Int. Conf. Adv. Comput. Commun. Syst. ICACCS 2021*, pp. 581–585, 2021, doi: 10.1109/ICACCS51430.2021.9442021
26. C. Shih, C. C. C. William, and Y. W. Chang, The Causes Analysis of Ischemic Stroke Transformation into Hemorrhagic Stroke Using PLS (Partial Least Square)-GA and Swarm Algorithm, *Proc. - Int. Comput. Softw. Appl. Conf.*, vol. 1, pp. 720–729, 2019, doi: 10.1109/COMPSAC.2019.00108
27. V. Kedia, S. R. Regmi, K. Jha, A. Bhatia, S. Dugar, and B. K. Shah, Time Efficient IOS Application for CardioVascular Disease Prediction Using Machine Learning, *Proc. - 5th Int. Conf. Comput. Methodol. Commun. ICCMC 2021*, no. Iccmc, pp. 869–874, 2021, doi: 10.1109/ICCMC51019.2021.9418453
28. M. Monteiro et al., Using Machine Learning to Improve the Prediction of Functional Outcome in Ischemic Stroke Patients, *IEEE/ACM Trans. Comput. Biol. Bioinforma.*, vol. 15, no. 6, pp. 1953–1959, 2018, doi: 10.1109/TCBB.2018.2811471
29. I. M. Chiu, W. H. Zeng, and C. H. R. Lin, Using Multiclass Machine Learning Model to Improve Outcome Prediction of Acute Ischemic Stroke Patients After Reperfusion Therapy, *Proc. - 2020 Int. Comput. Symp. ICS 2020*, vol. 9, pp. 225–231, 2020, doi: 10.1109/ICS51289.2020.00053

7

Multi-Modal Medical Image Fusion Using Laplacian Re-Decomposition

Kesana Mohana Lakshmi

CMR Technical Campus, Hyderabad, Telangana, India

Suneetha Rikhari

Mody University of Science and Technology, Lakshmangarh, Rajasthan, India

CONTENTS

7.1 Introduction

The aim of image fusion in medical imaging is to merge or fuse significant information about the same region of interest in the human body obtained from multiple sources with different sensors. The various imaging methods produce digital images which can be easily processed and enhanced using a computer vision system [1]. In multi-sensor fusion systems, the information obtained from different images is analyzed and complementary information from each image is merged to create a single image. The single image obtained will thus have more information than any of the individual ones. This multi-sensor imaging is very popular in remote sensing applications. In remote sensing, images are obtained from different sensors having different spectra of their own. Spectral image fusion helps researchers to combine the information from different spectra so that they can have all the relative spectral information in a single image. The advantage of image fusion is that both spatial and spectral information can be envisioned in a single image.

Most of the techniques developed so far are not well adapted to produce such fused images. Some of the conventional fusion methods have failed to protect the spectral information during the fusion process, resulting in spectral distortion [2–4].

Today, medical image fusion is producing promising results in clinical settings by combining medical images of the same region of interest. This fused information helps medical experts obtain a diagnosis and plan treatment accordingly. Traditionally, medical images have been obtained from different modalities like X-ray, magnetic resonance imaging (MRI), computed tomography (CT), positron emission tomography (PET), single photon emission computed tomography (SPECT) etc. When wide X-ray beams are passed through the subject, some of these rays reflect back upon hitting the hard tissues of the organs. The reflected rays are then imposed on a photographic film to generate an X-ray image. CT scan images are obtained by subjecting the patient to X-rays at multiple angles to get more detailed information in combination with a computer. CT image provides more detailed information than plain X-rays. CT images provide detailed hard tissue information as compared to soft tissue information. In MRI the patient is kept inside a high magnetic field, and the signal collected from an injection of radio waves is converted into a digital image with a computer. MRI is good at producing soft tissue information, but distorts hard tissue information. Using advanced fusion techniques, the hard and soft tissue information can be fused together into a single image. PET and SPECT modalities come into the category of nuclear medicine. They are used to study circulation and metabolic changes, and can easily detect clots in the heart or brain.

A variety of clinical image fusion techniques has recently been proposed. Because different modalities are used, images of varying intensities are obtained. Many fusion algorithms are thus designed in a multi-scale mode and are known as multi-scale transform fusion methods. This type of fusion is achieved in three phases—decomposition, fusion, and reconstruction. A hybrid multi-modal fusion of medical images is proposed in [5], using a combination of different transform techniques. Fusion of multi-modality images using discrete wavelet transform (DWT) is developed by [6]. A novel technique to fuse multi-resolution brain images was proposed by [7]. Another technique using new features in NSST domain is explained by [8]. Medical fusion using gradient transfer and latent low rank with MRI and CT images is proposed in [9]. A dictionary-learning medical image fusion based on entropy is discussed by [10]. In [11], non-subsampled contourlet transform is combined with PCA-DWT for fusion of medical images. IHS and log-Gabor transforms are used by [12, 13] for fusion of brain images with PET and MR modalities.

Many algorithms using rich weight assignment strategies and difficult decomposition methods to increase the efficiency of the fusion are proposed in the literature. Assigning weights and measuring the activity levels of the training network is a tedious task for a fusion algorithm. According to the literature, the conventional fusion method does not design a system in which the weights and measurement are done together. Two or more stages are used to separately calculate the weights and the activity levels, which may reduce the robustness of the algorithm. This problem is addressed in this chapter with the use of CNN and Laplacian pyramids. Initially a CNN is trained so that it can adaptively produce the weights while measuring the activity levels of the system. Since medical images are obtained from different modalities, a multi-scale image fusion is employed here using image pyramids. The use of CNN and pyramids makes for a more effective system as regards visual insight and human observation. Later, a similarity-based measurement which can alter fusion adaptively is applied to the source image coefficients.

7.2 Related Work

A multi-scale fusion method using CNN which can produce more appropriate and effective outputs was proposed in [14]. This model used a Siamese CNN which uses two similar branches of similar architecture with similar weights. The source images are applied as inputs to the two branches of the Siamese network. Three convolutional layers are present in each branch to produce the feature maps. These calculated feature maps are nothing but the activity level measurements of the network.

The calculated feature maps are combined and then applied as inputs to a fully connected equivalent convolutional layer network. This connection helps the fusion process to allow input images of any arbitrary size, which is an important part of the weight assignments. The coefficients generated in the network specify some important property of the source image patch considered at a particular location. For each coefficient generated there exists a focus important feature at the same location of the source image patch. Then, values are assigned to all the pixel weights corresponding to the image patch considered. The overlapped pixels are averaged and a focus map is obtained of similar size to the images of the source input. The final focus map is then generated using the weighted average rule. Two consistency verification methods are used along with the focus map to generate an output fused image. The practicability and the effectiveness of the fusion process using CNN are clearly explained in [14].

The central objective of the method employed in this work is to apply the CNN model used in [14] to medical images. For a number of reasons, the CNN technique alone may not produce efficient medical fusion. The main points to be considered before using CNN are:

1. The entire fusion process discussed in [14] is done in the spatial domain. The spatial domain approach may not produce a desirable fusion output because of the individual characteristics of the medical imaging modalities used. The transform domain approaches tend to produce fewer artefacts and robust fused images.

2. Moreover, the convolutional network is designed to have inputs with the same local structures. But applying inputs with same local structures is a difficult task in medical imaging because each modality will have its own intensity. Even though the same human organ is considered as the source input, the local similarity of the images is not possible because of the intensity variations of imaging modalities.

These two issues are resolved in the proposed work to generate a medical image fusion. The first issue is resolved by using a pyramid-based multi-scale model. The two input source images are first decomposed using a Laplacian pyramid [15] approach and the weight map generated by the CNN network is decomposed using a Gaussian pyramid approach. At each decomposition level the fusion operation is carried out.

The second problem is resolved by applying the local similarity fusion method to all the decomposed coefficients [15]. To avoid the loss of information, a weighted average method is applied to source image contents having high similarity. The weight map generated by CNN is used, rather than the coefficient-based weights. The 'choose max selection' mode is employed to preserve the vital features of the input images. After this, the activity level of the pixel measurement is calculated by direct measurement of decomposed coefficients' absolute values (at this stage the output of CNN is not used because of its ineffectiveness with medical images).

The method employed in [16] can therefore be effectively used to obtain fused medical images.

7.3 Proposed Methodology

7.3.1 The Convolutional Neural Network

The proposed methodology has employed the convolutional network depicted in Figure 7.1. This network is considered to be a Siamese network. It uses two similar branches with similar weights. One max-pooling layer and three convolutional layers are present in each of the branches (as in the network in [16]). The fully concatenated network layer is replaced by a simple model to increase computational speed and save memory.

The input patches p_1 and p_2 of size 16×16 are passed through a three-layer convolution network. Each branch after the third convolution layer generates 256 feature maps. These 256 feature maps are thus combined in the form of a two-dimensional vector by a softmax layer. This softmax layer is used to calculate the probability distribution function (PDF) of the two-layer classes. The PDF results in the normalized weight maps corresponding to the input patches applied. Each PDF value indicates the probability of weight assignment of each class. It is also considered that the addition of the two probabilities at the output is equal to 1. The PDF indicates the probability of weight assignment related to each class of the input patches.

Here, for training the network two source inputs are connected to each branch of the CNN. The patch p_1 is a taken from a high-resolution medical image and p_2 is said to be a blurred version of p_1. The patches p_1 and p_2 are assumed to be of size 16×16. Gaussian filtering and random sampling are used to create the training samples. The training process employed uses Caffe's [17] deep learning technique. The fully connected layer of the convolutional network is of fixed dimension. Since the dimension of the fully connected layer cannot be changed, the input to be connected to this layer must also be of fixed size. Hence,

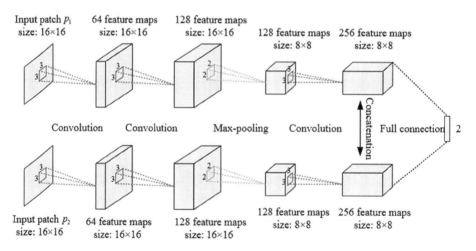

FIGURE 7.1
CNN training architecture.

the input and output data are assumed to be of fixed size. When the image is divided into patches of the same size there are overlapping patches. The overlapping patches may produce recurring calculations which are to be avoided to increase the speed of operation of the algorithm. This problem is addressed by fixing the size of each kernel to $8 \times 8 \times 512$. A two-dimensional vector prediction map is then generated after processing the source images (note that the source input images are of random size). Each vector gives a prediction value of each patch of the source at the equivalent location of the image. In each prediction, there are only two dimensions with their normalized sum to 1. Hence, the output obtained would be the weight of either of the input source patches. Lastly, the output of the CNN generates a two-dimensional weight vector obtained by assigning weights to each pixel of the input patch applied. The overlapped pixel weights are generated by calculating the average.

7.3.2 Pyramid Decomposition

A multi-scale representation called a pyramid is used to decompose the given image so that multi-resolution analysis can be done. In pyramid representation the input image is decomposed into several stacks in the shape of a pyramid. This decomposition is done by continuous smoothing and subsampling. In multi-resolution analysis, the pyramid representation of images is very helpful to analyze the image at each and every scale of resolution. The pyramids are mainly of two types: low-pass (LPP) and band-pass (BPP). Low-pass pyramids are formed by passing the image through a low-pass filter—generally a smoothing filter—after which subsampling is performed. The output of the first layer is a subsampled version of a smoothed image. Generally subsampling is done along the direction of each coordinate by a factor of two. A similar procedure is repeated with the output of the first layer, then the second layer, and so on for multiple times. Each step produces a subsampled image with decreased resolution. When represented pictorially, all the stacked-up layers look like a pyramid (Figure 7.2).

A band-pass pyramid is formed by considering the resolution at adjacent levels in the pyramid, taking the difference and doing interpolation [15, 18].

7.3.3 Gaussian Pyramid

A Gaussian pyramid is formed by a Gaussian average filter or Gaussian blur filter. Initially the corresponding images are blurred using a Gaussian average filter and then the filtered output is sampled down. Each lower level of the pyramid consists of an image with pixel values containing a Gaussian average corresponding to a pixel at the neighbourhood. This type of Gaussian pyramid representation is generally used for image texture analysis and characterization. Mathematically, the pyramid operation is described as follows. First the input image which is to be decomposed is blurred or low-pass filtered by convolving the input image with a Gaussian kernel. The parameter $\underline{\sigma}$ acts as the cut-off frequency of the low-pass filter.

Let the original input image be $I(x,y)$ and the Gaussian pyramid operation of image 'I' is

$$G_0(x,y) = I$$

$$G_{i+1}(x,y) = \text{REDUCE}(G_i(x,y))$$

FIGURE 7.2
The filtered images stacked one on top of the other to form a pyramid structure

The above operation is obtained by performing convolution of image I with the Gaussian filter. The mask of the Gaussian filter is taken in such a way that the pixel at the centre of the mask is considered to have more weight and its pixel neighbours chosen so that their sum is equal to 1. The convolving operation which produces Gaussian kernel is then determined by the equation:

$$w(r,c) = w(r)w(c)$$

where, $w(r) = \left(\dfrac{1}{4} - \dfrac{a}{2} \ \dfrac{1}{4} \ a \ \dfrac{1}{4} \ \dfrac{1}{4} - \dfrac{a}{2} \right)$ and $0.3 \le a \le 0.6$.

The expectation error $L_0(x,y)$ can be calculated as

$$L_i(x,y) = G_i(x,y) - EXPAND\big(G_{i+1}(x,y)\big)$$

The *EXPAND* operation is given by:

$$G_{i+1}(x,y) = 4 \sum_{m=-2}^{2} \sum_{n=-2}^{2} w(m,n) G_i\left(\frac{x-m}{2}, \frac{y-n}{2} \right) \tag{7.1}$$

If the values of $\left(\dfrac{x-m}{2}, \dfrac{y-n}{2} \right)$ are integers, then they are included in $G_{i+1}(x,y)$, otherwise they are discarded.

7.3.4 Laplacian Pyramid (LP)

The Laplacian pyramid is used to represent the image in a compact form. The difference between LP and Gaussian pyramid (GP) is that LP saves the difference images at each level. The last level is not the difference image. The calculation of differences at each level helps in the high-resolution reconstruction of an image. This type of representation and reconstruction is used in image compression.

Let G_o represent the input image to be decomposed. Initially, G_o is convolved with the low-pass filter or the Gaussian filter. Similar to LP after filtering, it is down-sampled by a factor of 2, which results in G_1.

Now, G_1 is up-sampled by adding zeros at each row and column. After up-sampling an interpolated low-pass filter is generated and is denoted as G_1'. Now the detailed image LP_o is generated by calculating the difference of G_o and G_1'.

$$\text{Detailed image } LP_o = G_o - G_1'$$

Hence, image compression can be effectively done by encoding LP_o and G_1 instead of directly encoding G_o because the low-pass filtered output G_1 can be represented at a reduced sampling rate and LP_o can be represented by a reduced number of bits. To attain more compression the procedure is repeated recursively on G_1 the same number of times as the size of the image. The final output will be the detail images $LP_o, LP_1, \ldots, LP_N$ and low-pass image G_N.

7.4 Fusion Method

The method proposed in this work is shown in Figure 7.3. The entire fusion process comprises the following stages:

- Producing weight map
- Laplacian pyramid decomposition
- Fusion of coefficients
- Pyramid reconstruction.

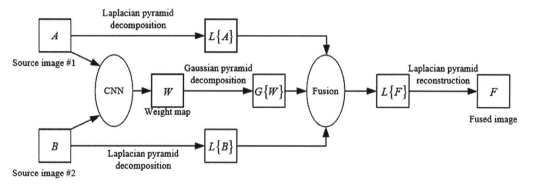

FIGURE 7.3
Medical image fusion method.

The weight map W is obtained from the CNN discussed above. The input images are labelled A and B respectively.

Next the pyramid decomposition is done by Laplacian pyramid. The Laplacian pyramids of the input source images A and B are represented as $L\{A\}^l$ and $L\{B\}^l$ where l is the lth level of decomposition. After decomposing the source images into pyramids, now the weight map is to be decomposed. A Gaussian pyramid is considered for decomposition of the weight map W. The Gaussian pyramid of the weight map is denoted $\{W\}$. The number of decomposition levels is restricted to $\lfloor \log_2 \min (H,W) \rfloor$, where H and W represent the spatial coordinates of the input images.

In the next stage, within a small window, the squaring of the coefficients is calculated and all the squares are added using Equation 7.2 at each decomposition level l. This results in a local energy map for each of the source images.

$$\sum_{A}^{l}(x,y) = \sum_m \sum_n L\{A\}^l \left(x+m,y+n\right)^2,$$

$$\sum_{B}^{l}(x,y) = \sum_m \sum_n L\{B\}^l \left(x+m,y+n\right)^2. \tag{7.2}$$

For similarity check, Equation 7.3 given below is used

$$M^l\left(x,y\right) = \frac{2\sum_m \sum_n L\{A\}^l \left(x+m,y+n\right) L\{B\}^l \left(x+m,y+n\right),}{E_A^l\left(x,y\right) + E_B^l\left(x,y\right)} \tag{7.3}$$

The similarity measure range is restricted to $[-1\ 1]$. If the calculated value is nearer to 1, it is supposed to have greater similarity. If it is closer to -1 there is zero similarity.

The fusion mode to be employed is determined based on the similarity measure check. If $(x, y) \geq t$, (where 't' is the threshold) the fusion mode with respect to the weight map W given by Equation 7.4 is considered.

$$L\{F\}(x,y) = G\{W\}^l(x,y) \cdot L\{A\}^l(x,y) + (1 - G\{W\}^l(x,y)) \cdot L\{B\}^l(x,y) \tag{7.4}$$

If $(x, y) < t$, Equation 7.1 is adopted for selecting the fusion mode. This is given by Equation 7.5.

$$L\{F\}^l\left(x,y\right) = \begin{cases} L\{A\}^l\left(x,y\right), & if\ E_A^l\left(x,y\right) \geq E_B^l\left(x,y\right) \\ L\{B\}^l\left(x,y\right), & if\ E_A^l\left(x,y\right) < E_B^l\left(x,y\right) \end{cases} \tag{7.5}$$

In the last step, reconstruction of the fused image F using Laplacian pyramid $\{F\}$ is done.

$$L\{F\}^l\left(x,y\right) = \begin{cases} G\{W\}^l(x,y) \cdot L\{A\}^l(x,y) + \left(1 - G\{W\}^l(x,y)\right) \cdot L\{B\}^l(x,y) & if\ M^l\left(x,y\right) \geq t \\ L\{A\}^l(x,y), & if\ M^l\left(x,y\right) < t\ \ \&\ \ E_A^l\left(x,y\right) \geq E_B^l\left(x,y\right) \\ L\{B\}^l(x,y), & if\ M^l\left(x,y\right) < t\ \ \&\ \ E_A^l\left(x,y\right) < E_B^l\left(x,y\right) \end{cases}$$

$$\tag{7.6}$$

The whole fusion process is conceded in a single equation defined by Equation 7.6.

7.5 Results

The experimental results are obtained using MATLAB. The input test images considered in this work are taken from publicly available sources (Figure 7.4). The main objective of a fusion algorithm is to achieve more informative data by combining the relevant data from the input data sets. The effectiveness of the fused output cannot be determined by visual inspection alone, so a statistical comparison of the fused output with the original data is made. Therefore, the proposed work demonstrates the output results both statistically and visually.

The proposed technique is also compared with DWT and stationary wavelet transform (SWT) and the image output results are depicted in Figures 7.5–7.7 with respect to the input data sets 1 & 2. The metric with a robust value is shown in bold. For statistical comparison of the results the following metrics are considered: peak signal-to-noise ratio

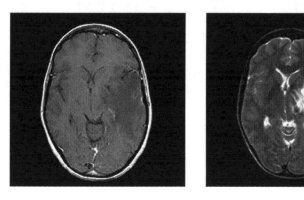

FIGURE 7.4
Test images (a) dataset 1 (MR-Gad & CT) (b) dataset 2 (MR-T2 & CT) (c) dataset 3 (MR-Gad and MR-T2).

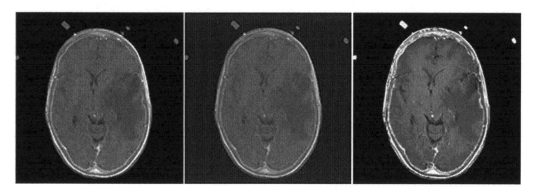

FIGURE 7.5
Fused images of dataset 1: (a) DWT (b) SWT (c) proposed method.

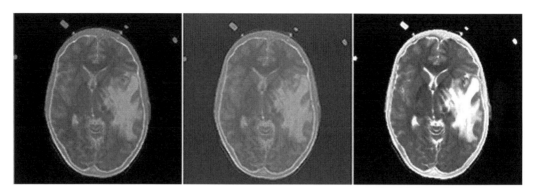

FIGURE 7.6
Fused images of dataset 2: (a) DWT (b) SWT (c) proposed method.

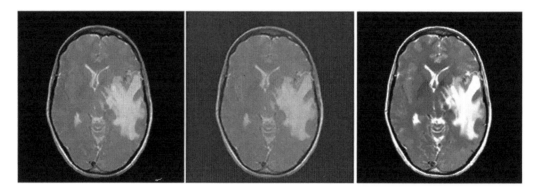

FIGURE 7.7
Fused images of dataset 3: (a) DWT (b) SWT (c) proposed method.

TABLE 7.1

Statistical Comparison for Input Set 1

Technique	PSNR (in dB)	RMSE	CC	SSIM	Entropy
SWT	68.95	0.0909	0.94	0.988	1.12
DWT	68.69	0.093	0.944	0.98	1.11
Proposed method	**90.51**	**0.00047**	**1**	**1**	**4.37**

(PSNR), structural similarity index measurement (SSIM), correlation coefficient (CC), root mean square error (RMSE) and entropy (E).

The optical quality of all the output images is better compared to SWT and DWT. As Figures 7.5(c), 7.6(c) and 7.7(c) show, the proposed method is able to produce high contrast and edge-preserved images. The SWT and DWT output images have less edge information and contrast. Moreover, the proposed algorithm, having high entropy, is able to preserve the texture information. The results are verified on three different dat sets and are shown in Figures 7.5–7.7.

Table 7.1 depicts the statistical assessment of the proposed technique with respect to SWT and DWT, using dataset 1 and considering the following metrics: PSNR, RMSE, CC, entropy

TABLE 7.2

Statistical Comparison for Input Set 2

Technique	PSNR (in dB)	RMSE	CC	SSIM	Entropy
SWT	63.56	0.169	0.867	0.975	1.03
DWT	63.60	0.16	0.871	0.975	1.02
Proposed method	**86.60**	**0.0007**	**1**	**1**	**4.76**

TABLE 7.3

Statistical Comparison for Input Set 3

Technique	PSNR (in dB)	RMSE	CC	SSIM	Entropy
SWT	65.05	0.142	0.885	0.979	1.009
DWT	65.02	0.14	0.89	0.979	0.99
Proposed method	**88.19**	**0**	**1**	**1**	**4.90**

and SSIM, With the method that has produced the best results highlighted in bold. As can be seen in Table 7.1, the proposed method produced better results than SWT and DWT. Similarly, Tables 7.2 and 7.3 depict the values of the metrics for dataset 2 and dataset 3.

7.6 Conclusion

The proposed work demonstrates an efficient fusion method for medical images using CNN and multi-scale methodology. The convolutional neural network is a two-branch Siamese network with similar branches and similar weights. The output of the CNN generates a weight vector map with the direct mapping of the pixel activity of source images. The key benefit of the suggested method is that it can adaptively do the weight assignment and pixel activity level measurement together via the same learning network. The values of the statistical parameters are different for different data sets. The PSNR value of the proposed method for data set 1 is high compared to data set 2 and data set 3 (Figure 7.8).

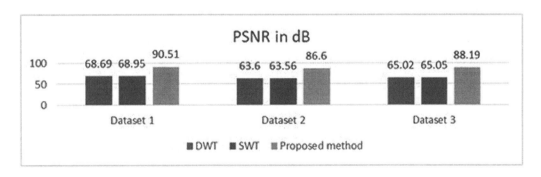

FIGURE 7.8

Performance analysis of PSNR for dataset 1, dataset 2 and dataset 3 with existing and proposed fusion methodologies.

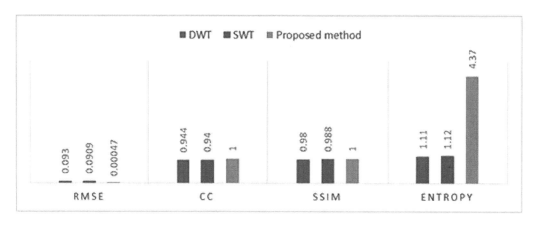

FIGURE 7.9
Performance analysis of RMSE, CC, SSIM and entropy for dataset 1 with existing and proposed fusion methodologies.

However, compared to SWT and DWT, the peak signal-to-noise ratio is greater for all the data sets of the proposed method. As shown in Figure 7.9, the proposed method is able to generate high entropy which has resulted in texture preservation while fusion occurs. In this research work, MR and CT medical image fusion has been performed. The work can be extended to other medical imaging modalities like SPECT, PET and 3- D images. The work is done using noise-free images. In real-time applications the images may be corrupted by noise. Hence, there is scope for developing new algorithms which can both remove the noise and perform fusion simultaneously.

References

1. Wei Tan Prayag Tiwari Hari Pandey Catarina Moreira, and Amit Jaiswal, 2020. Multimodal medical image fusion algorithm in the era of big data. *Neural Computing and Applications*, 1–21.
2. Bing Huang Feng Yang Mengxiao Yin Xiaoying Mo, and Cheng Zhong, 2020. A review of multimodal medical image fusion techniques. *Computational and Mathematical Methods in Medicine*, 70, pp. 625–631.
3. B. Meher, S. Agrawal, R. Panda, and A. Abraham, 2019. A survey on region based image fusion methods, *Information Fusion*, 48, pp. 119–132.
4. B. Rajalingam, and R. Priya, 2017. Multimodality medical image fusion based on hybrid fusion techniques. *International Journal of Engineering and Manufacturing Science*, 7(1), pp. 2249–3115.
5. Rajalingam Balakrishnan Rainy Priya, and R. Bhavani, 2019. Hybrid multimodal medical image fusion using combination of transform techniques for disease analysis. *Procedia Computer Science*, 152, pp. 150–157.
6. V. Bhavana, and H. K. Krishnappa, 2015. Multi-modality medical image fusion using discrete wavelet transform, *Proceedings of 4th International Conference on Eco-friendly Computing and Communication Systems*, 70, pp. 625–631.
7. Leena Chandrashekar, and A. Sreedevi, 2017. A novel technique for fusing multimodal and multiresolution brain images, *Procedia Computer Science*, 115, pp. 541–554.
8. P. Ganasala, and V. Kumar, 2014. Multimodality medical image fusion based on new features in nsst domain, Biomedical Engineering Letters, 4(4), pp. 414–424.

9. L. Meng, X. Guo, and H. Li, 2019. MRI/CT fusion based on latent low rank representation and gradient transfer, *Biomedical Signal Processing and Control*, 53, p. 101536.

10. G. Qi, J. Wang, Q. Zhang, F. Zeng, and Z. Zhu, 2017. An integrated dictionary-learning entropy-based medical image fusion framework, *Future Internet*, 9(4), p. 61.

11. S. Madanala, and K. Jhansi Rani, (2016). PCA-DWT based medical image fusion using non sub-sampled contourlet transform, *Proceedings of International Conference on Signal Processing, Communication, Power and Embedded System (SCOPES)*, October, Paralakhemundi, India.

12. C.-I. Chen, 2017. Fusion of PET and MR brain images based on IHS and Log-Gabor transforms, *IEEE Sensors Journal*, 17(21), pp. 6995–7010.

13. M. S. Dilmaghani, S. Daneshvar, and M. Dousty, 2017, A new MRI and PET image fusion algorithm based on BEMD and IHS methods, *Proceedings of Iranian Conference on Electrical Engineering (ICEE)*, May, Tehran, Iran.

14. Yu Liu, Xun Chen, Juan Cheng, and Hu Peng, 2017, A medical image fusion method based on convolutional neural networks, *Proceedings of 20th International Conference on Information Fusion*, July 10–13, Xi'an, China.

15. K Koteswara Rao, and K Veera Swamy, 2021. Multimodal medical image fusion using Laplacian redecomposition, *IOP Publishing*, 1070. 012080. pp. 1–13.

16. Z. Wang, and Y. Ma, 2008. Medical image fusion using m-PCNN, *Information Fusion*, 9, pp. 176–185.

17. Y. Jia, E. Shelhamer, J. Donahue, S. Karayev, J. Long, R. Girshick, S. Guadarrama, and T. Darrell, 2014, Caffe: Convolutional Architecture for Fast Feature Embedding. *Proceedings in ACM International Conference on Multimedia*, Association for Computing Machinery, New York, NY, USA, pp. 675–678.

18. J. Du, W. Li, B. Xiao, and Q. Nawaz, 2016. Union Laplacian pyramid with multiple features for medical image fusion, *Neurocomputing*, 194, pp. 326–339.

8

Blockchain Technology-Enabled Healthcare IoT to Increase Security and Privacy Using Fog Computing

S. Gunasekaran

Ahalia School of Engineering and Technology, Palakkad, Kerala, India

S. Shanmugam

Saveetha Engineering College, Chennai, India

D. Palanivel Rajan

CMR Engineering College, Hyderabad, India

P. Rontala

School of Management IT and Governance, University of KwaZulu-Natal Westville, Durban, South Africa

CONTENTS

DOI: 10.1201/9781003240310-8

8.1 Introduction

The parallel development of blockchain technology and internet of things (IoT) technology is enabling the healthcare sector to create a more patient-friendly and transparent system in a secure and distributed way. The development of cloud computing and fog computing has brought the medical environment into the digital age. Electronic health records (EHRs) maintain patient health information with various parameters recorded in hospital during treatment and for further treatment, care, or billing information [1]. Remote patient monitoring (RPM) using IoT and the cloud allows patients to enjoy the inherent benefits of care at home. Patients can maintain contact with healthcare professionals as necessary, reducing expense and improving treatment [2]. Using IoT technologies, the medical manufacturing sector can make supply-chain management transparent and provide an active tracking system. Medical data like genomic information can be digitalized and accessed anywhere and anytime using cloud computing. However, one of the major issues for all these developments is security. Blockchain technology resolves all the security issues and provides the most advanced, distributed, transparent, and encrypted security services at all levels.

Modern healthcare faces numerous problems—fragmented data, disparate workflow tools, and delayed communications—due to a lack of interoperability. There is a need for a fragmented system that allows secure admission to complete and recognized longitudinal medical records with manipulations. Blockchain-based system scan be used to maintain medical records on a decentralized platform that initiates and empowers usage with incredible speed while also providing a transparent and secure environment—a distributed system that provides a ledger facility for securely storing records as well as interoperability for supporting applications [3]. Blockchain technology eliminates the need for a centralized authority to verify the integrity and ownership of information, and also to arbitrate transactions and exchange of digital resources, as well as enabling security in undisclosed transactions and agreements directly between the interacting nodes. Its key properties, including decentralization, immutability, and transparency, can address urgent health care issues such as incompletepeer-to-peer files and difficulty accessing patient health information.

Cost reduction is a top priority for many hospitals and medical professionals because payers, including private insurers, have begun to link payment of claims to quality and cost achievement [4]. Some companies have reduced costs by examining their revenue billing cycle and service operations, while others have looked at their healthcare supply-chain management. Pharmaceutical companies, hospitals, insurance companies, different purchasing organizations, and regulatory agencies are among the stakeholders involved in the supply chain. However, when hospitals, clinical and pharmaceutical companies work together to improve the overall efficiency of the supply chain, they see a significant cost reduction.

The increasing use of genomic data in medical research had led to the creation of large-scale genomic big-data repositories. The combination of blockchain technology and cloud computing technologies is causing a revolution in the field of genomic data management.

With blockchain technology, endusers will be able to sell their valuable genomic data on the most secure and transparent platform possible. Manny healthcare organizations, such as Shivom, Opal/Enigma, Nebula, and Health Nexus, are providing these services using a secure blockchain and cloud-based infrastructure. This article will discuss the sharing of clinical data using the FHIR chain [5] proposed by P. Zhang et al.

The chapter is organized in four sections. Section 8.2 looks at blockchain with healthcare IoT and Ethereum. Blockchain technology in healthcare supply-chain management and related companies is explored in Section 8.3. Section 8.4 discusses secure genomic data transactions using blockchain technology. Conclusions are provided in the final section.

8.2 Blockchain with Healthcare IoT and Ethereum

Wearable technology and the IoT have led to a massive increase in the amount of healthcare-related data generated. These data are of great value to everyone involved in the healthcare ecosystem. Currently, the vast majority of this sort of data is segregated and scattered across private or proprietary databases. EHRs are dynamic, patient-centred records that make data available to authorized users instantly and securely. The EHR records all clinical and diagnostic information about a patient, as well as instructions for dispensing medications and scheduling forthcoming treatments. The use of EHRs increases patient involvement, enhances care coordination, improves diagnoses and patient outcomes, and saves money.

For patients with IoT-enabled devices connected to the internet using fog computing, data is transferred and distributed securely. The fog makes a close connection with end users' IoT devices to alleviate the impact of such unsupported features of IoT applications as location awareness, low latency and geographic distribution. Fog computing also reduces data offload to the cloud, which adversely affects response time [6] (Figure 8.1).

The exchange of health-related data is vulnerable even after security and privacy concerns have been addressed, and there is a lack of data authenticity tracing. A reliable option for overcoming these problems is decentralized and consensus-driven distributed ledger technologies (DLTs), such as IOTA Tangle, blockchain and Ethereum [7]. Health-related environmental data from stationary sensors and activity data from wearable sensors are very useful in commercial and research healthcare applications. All connected parties, including device patients, users, researchers, and companies, will benefit from this type of proper sharing of health data, which will progress in the public health care system.

In addition to complex data protection regulations, another major impediment to the free flow of big data is the high cost of transferring sensitive granular data in real time because of intermediary fees. Another constraint is the exclusion of guaranteed data validity and audit trails. Man-in-the-middle attacks and data manipulation will frequently target traditional data transmission protocols and databases. A distributed ledger is a database regularly updated by a node in a peer-to-peer network using an agreed procedure. Because all peers contribute to the integrity of the database, this consensus technology reduces the need for central administration [8].

8.2.1 Distributed Ledger Technologies and Blockchain

A distributed ledger is a distributed database brought up to date by an agreement protocol that will run for the peer-to-peer network nodes. This agreement protocol removes the

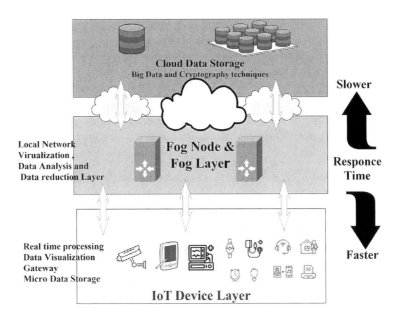

FIGURE 8.1
Communications between the three layers of fog computing.

need for central administration because all peers are needed to contribute to the integrity of the database [9]. Blockchain is one of the most widely used DLTs, and has become better known in recent years owing primarily to crypto-currencies in the financial sector. The risk of data breaches can be reduced by setting a threshold value for encrypted data for a public key infrastructure, where more users must collaborate with decrypted data and asymmetric cryptography, used to make messages authentically with the system participants [10].

8.2.2 Limitations of Blockchain

New decentralized applications such as smart contracts and cryptographic currencies are enabled using DLT-based distributed agreement protocols. Over the last six years, the rise and success of Bitcoin and Ethereum have established blockchain technology which has an inherent value. In certain situations, when using block-based protocols like blockchain and Ethereum, there are some problems that render a generic IoT data-sharing platform undesirable [8]. Mining nodes on the blockchain network may consume a lot of energy while they are updating the ledger and executing complicated algorithms. While authorized networks are more efficient, public networks require a significant amount of energy to maintain their operating status. When the blockchain continues to grow in terms of transactions and nodes, the entire network becomes slower. For commercial blockchains, it is critical for the network to be both fast and safe at the same time. Double-spending and DDS attacks are some of the hacking difficulties connected with the blockchain network, which may be avoided by utilizing various consensus algorithms, such as Proof-of-Stake, Proof-of-Work, and so on. Transactions using blockchain money are limited or illegal in various countries. It is difficult for users from these nations to keep track of their wallets, and because most users are not technologically literate, they make mistakes when selling medical information and genetic data.

8.3 Supply-Chain Management in Healthcare: Blockchain Technology

8.3.1 Supply-Chain Management (SCM)

SCM identifies the flow of commodities and services that transform raw materials into finished products. In SCM, the end customer is regarded as a supplied component, along with all the various goods and products. SCM is responsible for managing the flow of raw materials through the entire network of business processes and activities such as purchasing, distribution, manufacturing and delivery of finished goods [11]. One of the goals of a good SCM provider is to ensure that the correct product is supplied to the right customer at the right time and place, at the right cost.

In every business process, procurement is crucial, from business needs to IT needs. Today we are more reliant on suppliers than in the past. Despite increasing the requirement to manage supplier relations, information, contracts, and more, the requirements to adhere to regulations persist [12].

A good SCM system will help achieve these objectives:

- Managing contractual obligations to ensure a constant supply and prevent service disruptions, such as interruptions to deliveries.

- Empowering suppliers to enhance more significant components for synergistic growth with multiple lines of business.

- Providing procurement management to ensure that proper suppliers are used, and saving money.

- Acting in accordance with compliance requirements within the company as well as in the industry.

- Ensuring that only one overall supplier view is in place, along withcollecting valuable procurement analytics.

8.3.2 Healthcare Supply-Chain Management

The healthcare supply-chain process begins with medical product manufacturers dispatching the materials to the distribution centre. In the majority of circumstances, hospitals will directly acquire products from manufacturers or distributors. If the product is unusual, the hospital can contract with a group purchasing organization (GPO) to handle the transaction. Both regulatory authorities and payers are critical in healthcare supply-chain management [13]. A medical resource may be licensed for consumer use only after being assessed by regulatory bodies and determined to be safe and effective for use by specific patients (Figure 8.2).

Significant challenges in healthcare supply-chain management are: (i) shortage of supplies impacting on the quality of care; (ii) lack of automation and manual processes; (iii) flaws in health IT systems; and (iv) costs that are not obvious.

The healthcare industry relies heavily on data. Everything from drug supply chains to health records is currently managed using traditional computer and documentation systems. Inconsistencies among the parties participating in the data management system prolongs and increases the cost and reliability of the process [14]. Blockchain in healthcare improves the security of patients' EHRs, drug authenticity and supply-chain tracking issues, and enables secure interoperability among healthcare organizations [15].

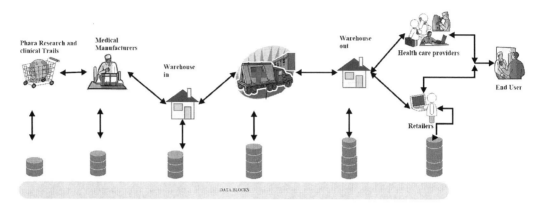

FIGURE 8.2
Overview of blockchain with supply-chain management.

While blockchain adoption in healthcare has been slow, a significant shift is underway. One in every five medical and healthcare organizations used blockchain by 2020, according to IDC. By 2022, blockchain will have been used for business purposes in 55% of all healthcare applications. Every year globally there are approximately 30 billion healthcare transactions, 50% of which are faxes [16]. These have a staggering cost of $250 billion to the industry.

Moreover, 63% of hospitals do not receive data on time or receive insufficient information via reimbursement letters considering the high costs. Every year, there are 20,000 deaths due to administrative errors, 80% of which are caused by miscommunication of patient-related data [15]. Manual paperwork is another issue, with each provider needing to fill in nearly 20,000 forms annually [17], at an average cost of $20. Blockchains are distributed ledgers that are immutable, transparent, and easy to access from everywhere, and these properties will be crucial to the reformation of the present healthcare sector.

Traceability and reliability are critical in this industry. Any drug company wishing to register products on the blockchain must be deemed reliable by the private healthcare blockchain's administrators. Once approved, they can store all data related to drugs on the blockchain. Every block on the blockchain will be digitally signed and its specification will be immutable, so that no party to drug transactions, from manufacturer to retailer, can tamper with the data or add counterfeit pharmaceuticals and drugs [18].

Medication can be tracked in real time on its journey using data stored on the blockchain [19]. QR codes on packaging can assist with product authentication.

8.3.3 Pharmaceutical Supply-Chain Management

Pharmaceutical supply chains are complex [20], with stakeholders from multiple disciplines making demands on them and with many criteria necessary for their input. The supply chain must ensure that data is verifiable and that multiple parties can interact with it, including international and national verification systems, transport systems, and regulatory bodies [21].

There are more stakeholders with more opportunities to change and influence the quality of the data. Tracking and validating correct information while mitigating human error and missing documentation is a difficult task, especially as interactions are often time sensitive [22]. Manufacturers, distributors, wholesalers, and pharmacies frequently have

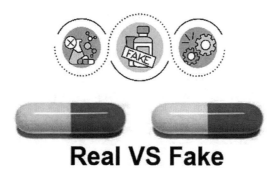

FIGURE 8.3
Identification of real vs fake drugs.

difficulty ascertaining the authenticity and quality of medication while it is being shipped and distributed (Figure 8.3).

The supply chain can become a holistic, accurate, audited, and secure process, with drugs trackable all the way to the patient [5] and identification processes that involve scanning barcodes and recording them onto a blockchain ledger system [23]. All deliveries can be tracked, with biometric identification procedures used to trace the delivery driver. Checkpoints that include drug measures can also be tracked using several devices [24].

8.3.4 Blockchain-Based Healthcare Companies

8.3.4.1 Akiri

Akiri [25], devised by the American Medical Association (AMA), is a cloud platform service used for securely exchanging data via a standardized system of codes. It provides the most secure and independent transfer of medical data without retaining any data.

8.3.4.2 BurstIQ

BurstIQ [26] has solved the basic difficulty that has prevented blockchain from being used in healthcare, securely and privately storing vast, complicated, and heterogeneous data on-chain. BurstChain™manages data rights end-to-end, ensuring an unbreakable chain of custody, granular ownership and revocable consent, and superior security.

8.3.4.3 Factom

Factom [27] is a blockchain-based distributed publication technology that uses immutable and independently verified record systems. It is designed to maintain the privacy of sensitive data and avoids the requirement for trusted intermediaries. While earlier public blockchains (such as Bitcoin) used a distributed ledger design to connect relevant entries chronologically, Factom used a distributed ledger design to connect related entries sequentially for improved storage and retrieval. Entries may include any information but they are not designed to store private data. Data entering the entry block is hashed and saved in distributed hash tables before being written, and the actual data is kept in distributed hash tables and swapped peer to peer. By securing each directory block using Bitcoin's cryptographic anchor entries, the Factom blockchain makes that information far safer.

Interoperability and third-party security are enabled by being anchored into other public blockchains. Factom's public blockchain is based on federated servers identified by their contributions and performance. Anyone can see the posts, and everyone can submit written requests for a charge based on the size of the chain and the amount of information in the message. Anyone can host a follower node or construct a private network for development and production purposes because the Factom protocol is open.

8.3.4.4 MedicalChain [27]

The main goal is to give the patient control over their medical data by sharing the most comprehensive version of their record with every organization in their medical network. Patient records that are fragmented and siloed cause inefficiencies and inaccuracies throughout the healthcare system. MedicalChain employs blockchain technology to securely manage health records to provide a collaborative, intelligent approach to healthcare.

8.3.4.5 ProCredEx [28]

Researchers at ProCredEx are enthusiastic about credentials data and the benefits it can bring to healthcare when organized, analyzed, and shared using cutting-edge technology. ProCredEx has merged decades of expertise in healthcare credentials with a premier distributed ledger and advanced analytics technologies. Long-standing credentialing practices are becoming more efficient, and this complex dataset is transformed into insights that improve patient safety, quality of care, and decision-making. Avaneer, SimplyVitalHealthee, RoboMed, Embleema, and Chronicled are some of the other companies.

8.4 Genomic Data

A genome is a recipe for the development of an organism. Genomes are stored in cells as deoxyribonucleic acid (DNA), an extraordinary molecule that creates and nurtures organisms according to an encoded pattern. For example, the human genome contains about 23,000 recipes, each of which contains the information needed to construct one or more proteins. These recipes or functional units of DNA are referred to as genes.

Genetics [11] is the branch of science that is concerned with the structure and function of the genetic code. One percent of DNA is composed of genes that code for proteins; the remainder is non-coding DNA, sometimes known as junk DNA. In the non-coding DNA sequences, DNA sequences provide information for synthesizing particular types of RNA molecules (RNA is the chemical cousin of DNA). In the making of protein, specialized RNA molecules called transfer RNAs (tRNAs) and ribosomal RNAs (rRNAs) are created from non-coding DNA. These molecules help in the construction of amino-acid chains, which constitute a protein. Telomeres protect chromosomal ends from destruction during genetic material replication.

Researchers at the Human Genome Project, which has had a significant impact on life sciences research, needed to devise innovative ways to organize and distribute massive amounts of data, which triggered a wave of studies culminating in 2001. By looking at the entire genome rather than just a single gene, researchers established that the human genome has two copies of any given piece of DNA. We know that large segments of DNA in the genome can be replicated or deleted. Researchers in Canada have compiled a data-base to further understanding of how genomics might benefit human health.

Among the most significant problems facing the biomedical industry are personal genetic data sharing and genetic applications, which raise complicated privacy concerns. The major challenges in privacy protection technologies for genomic data are transferring and accessing person-specific DNA that is not associated with an explicit identity (i.e., name, social security number, etc.). The HIPAA Privacy Rule, established by the U.S. Department of Health and Human Services,[1] establishes national standards for protecting individual medical records and other personally identifiable health information. It applies to health plans, health care clearing-houses, and health care providers who conduct certain health care transactions electronically.

Today, the genomes of different animals, plants, and bugs have been read by comparing genomes. The ever-growing amount of publicly available data allows us to study how diverse species will act today and throughout evolutionary history. Everyone has access to genomic resources and this free access to such rich knowledge enables many beneficial applications that influence our lives, our health, and our environment [29].

8.4.1 Empowering Genomic Blockchain Technology

Conger and Kanungo (1988) propose a four-stage model to describe the process of empow-erment, which consists of the various phases of psychological empowerment, antecedent conditions, and consequent behaviours. This concept has been extensively used in order to better understand how to impact subordinates' emotional states through empowering strategies, thereby influencing behaviour. It has also been used to explore the relationship between information systems (Figure 8.4).

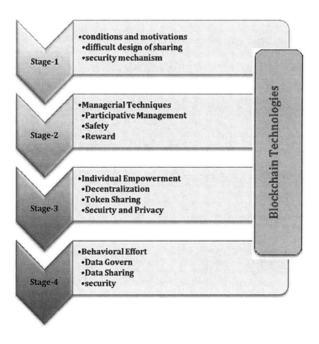

FIGURE 8.4
Stages of an empowerment process.

Conger and Kanungo's (1988) four stages of an empowerment process [24] are:

Stage 1   *Conditions and motivations.* While several factors inhibit the growth of genomic big data, three basic primary issues are ownership of data, problems with sharing the data among the healthcare community, and providing most secure access rights.

Stage 2   *Managerial techniques.* Due to the unique properties of blockchain, such as decentralized, smart bonding, anti-tampering, traceability, and encryption algorithms, anyone on the blockchain platform can participate in the data management process. Additionally, participants can potentially earn incentives in the form of tokens as a result of data exchange. The incentives given in the form of tokens might be interpreted as a type of competence-based reward. A greater degree of data security and privacy can be ensured with the implementation of blockchain technology.

Stage 3   *Empowered individuals.* Blockchain technology offers a unique and useful concept for a genomic big-data platform. Data ownership is allocated to the individual, and authority delegated to them. The token system may be of assistance in the data-sharing method, and security and privacy are better protected as a result.

Stage 4   *Behavioural consequences.* In the long run, platform users will change their behaviour as a result of the platform's characteristics. Individuals will be provided with greater control over their personal data, and this process will inspire people and institutions to be more active in exchanging ideas. Moreover, with increased privacy and security measures, the public will be invited to use this platform more regularly. It helps individuals to perceive things in new ways and promotes behavioural changes.

8.4.2 Challenges of Genomics Big-Data Platforms

One of the most important functions of the genomics big-data platform system is to collect and store person-specific genetic data for the purpose of implementation and distribution. Poor data quality, information islands, manipulation distortions, missing records, individual privacy leaks, and grey data transactions, among other factors, commonly obstruct the collection and sharing of genetic data [24]. The most significant impediments to the creation of a genomics big-data system can be classified into three categories: ownership of data, techniques of data transfer, and data protection [30]. It is likely that in the near future, worldwide standards will be implemented to streamline data ownership guidelines and procedures.

8.4.3 Genomic Data Platform Architecture

X.-L. Jin proposed the LifeCODE.ai [31] platform, which aimed to integrate genetic data with phenotypic health data, maintaining maximum data quality and confidentiality, on a platform where various services and data can be exchanged, shared, and utilized by everyone. The platform's architecture is made up of four levels (Figure 8.5). A fundamental infrastructure, data, blockchain, and user interface layer must be in place before adding more layers. Business services and related APIs are found in the interface layer, and the blockchain layer, which benefits from encryption, decentralization, traceability, and

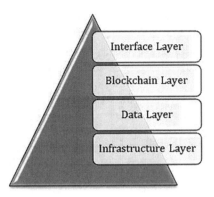

FIGURE 8.5
LifeCODE.ai layered architecture.

anti-tampering measures, is the backbone of the platform. It also incorporates a CAM-brain engine, a key management system, and smart contracts. Data security is further protected by searchable encryption and trustworthy data storage, both included in the data layer. Access control network, basic security services, and other services may be found at the infrastructure layer. To provide reliable services at the front end, the blockchain data and infrastructure layers are utilized together in the interface layer, allowing sophisticated algorithms and safe storage to take place at the back end.

8.4.4 Case Study of Genomic Data Sharing in LifeCODE.ai

Users of LifeCODE.ai can connect with trade data and data investors in order to obtain data for their own purposes. LifeCODE.ai leverages the blockchain's smart-contract and anti-tampering capabilities, as well as other technologies, to facilitate data sharing via "tokens". In the case of diabetes, for example, a research institution seeking 100 pieces of data from diabetic patients for medication development may search the platform using the term "diabetes". They can obtain anonymized contact information for the right individuals and make a token offer to them. Individuals can access services from medical and insurance organizations using the tokens and they have the option of accepting or rejecting a contract. By exchanging tokens for data, users may be more inclined to contribute their personal genetic data. During data exchange, smart-contract code enables, validates, and enforces contract or business negotiation. The anti-tampering feature contributes to data accessibility and integrity protection, which ensures the proper operation of smart contracts and their usefulness.

The following four phases constitute the token system:

1. The data holder's information is imported into the GOR architecture, which is specifically designed for the storage and analysis of genetic data. The GOR architecture, GOR pipe, clinical sequence analyzer, and API risk engine are able to detect probable genetic-clinical links and provide that information to the clinical team [32].

2. LifeCODE blockchain [31] collects data from various reliable, authorized sources validated by the permission mode. The data encryption, asymmetric encryption techniques and tag-based fingerprint extraction are implemented and a keyword search is established by the data label.

3. LifeCODE.ai indexes and categorizes the phenotypic data, combining it in a range of ways utilizing protocols and adapters, depending on logic and experience.

4. The process of data sharing is backed by secure data transfer, which is a type of encryption. LiveCode is an authorized blockchain based on the corporate application-oriented Ethereum by JP Morgan Quorum. Permissioned blockchains enable restriction of some actions to designated members.

8.4.5 Genomic Blockchain Technology

8.4.5.1 Encrypgen [21]

Encrypgen is a decentralized, democratic, and dis-intermediate platform for sharing genetic data utilizing blockchain technology. Users can maintain their privacy while earning money by selling access to various sections of the genome. Encrypgen secures all genomes in the cloud and on private servers, while a bespoke blockchain maintains lightweight metadata about off-chain files and serves as a transaction audit trail. An equal amount of new DNA tokens will be added to the Gene-Chain platform wallet or an Ethereum wallet such as My Ether Wallet or Metamask. This can publish the contract address where the user can add them once the exchange is complete.

8.4.5.2 Health Nexus

Health Nexus utilizes distributed ledger technology (blockchain) for the transfer, payment, and storage of health data [21]. The project aims to create a blockchain system that is secure, compliant, and capable of handling strict healthcare regulations by implementing more robust data encryption, integrity, and validation protocols for mining companies. To maintain the integrity and security of the database, and to manage executive validation, an organizational governance system and an upgrade to the data integrity and security system are installed.

Medical-quality metrics, transparency, insurance payments and reimbursements, pharmaceutical tracking, data access and sharing, and digital insurance are all provided as application services in this platform via the use of HIPAA-compliant protocols. Data storage is secured using a distributed hash table to manage a specially modified Kademlia table.

8.4.5.3 Nebula

Nebula, another genomic data-exchange system using blockchain technology in decentralized, fast, and secure transactions, uses the homomorphism for data encryption and integrates the Arvados [33] bioinformatics platform[2] with the Exonum [34] blockchain framework.[3]

Arvados provides distributed computation of data storage. Its two key services are Keep and Crunch. Keep is a distributed content-addressable storage solution that stores large amounts of big genomic data. It provides fast data access and simple data management. Crunch enables parallel creation and execution of data analysis pipelines as well as reproducible results. It helps deal with big-data issues, restrictions on the use of data, and privacy issues. Using homomorphic data encryption allows for privacy-preserving queries on genomic data, which is used to protect data privacy by allowing investigators to query the entire database and find data of interest without compromising data privacy.

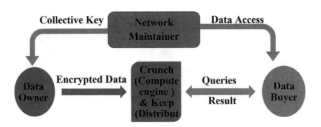

FIGURE 8.6
Nebula GENOMIC data exchange.

The Nebula network consists of four players: data owners, network maintainers, data buyers, and storage providers. Data owners can provide genetic data in encrypted form. They handle their own data and wallet payments. All validator nodes handle encrypted key shares, and keep track of data and Crunch computations. The auditor nodes keep a copy of the metadata to identify data in Keep, check data integrity, and monitor access permissions. Data buyers can access the homomorphically encoded data using smart contracts and run analytical pipelines with Crunch. The storage provider makes use of a cloud-based infrastructure (Figure 8.6).

8.4.5.4 Opal/Enigma

OPAL stands for "open algorithms" [6]. Enigma is used for secure distributed and encrypted transactions using a distributed data repository architecture with the help of a peer-to-peer network. Data is encrypted at its repository to avoid raw data leaks. Data operations are defined by legally binding smart contracts authorized by assigned digital identity credentials. A clinical trial infrastructure comprising standards for consent, patient registration, trial management, and data management will be used to make OPAL/Enigma a reality (Figure 8.7).

In OPAL/Enigma's three-layer architecture, the middle or crypto infrastructure layer provides secret sharing and multi-part computation process for sharing the information among common groups with security for individual data items. In the decentralized P2P network, nodes keep relevant data items which are distributed arbitrarily. With a decentralized P2P node network, the data owner no longer has to maintain a centralized database.

8.4.5.5 Shivom

The Shivom project [21] is an attempt to make genomics accessible to everybody. It aims to provide an environment where customers may learn and purchase genomic services on the blockchain. The value of this business lies in the empowerment it provides individuals by giving them complete control over their genetic data, which could revolutionize healthcare delivery.

OMX tokens are used to fuel the Shivom ecosystem, and these tokens may be used to buy products and services on the platform across international borders. It is as if everything in the Shivom ecosystem can be "tokenized". The Shivom Token is a form of digital currency representing a contract. Therefore, it can be used to recognize participation or data sharing in different countries around the world and allow participants to gain access to genome sequencing kits, genetic reports, and applications in the Shivom marketplace.

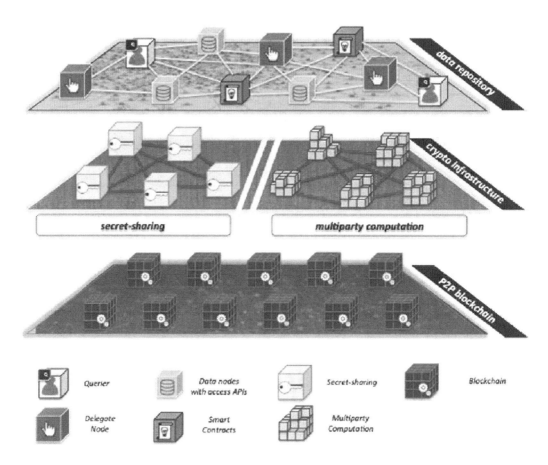

FIGURE 8.7
Layers of OPAL/Enigma.

Shivomhas developed a genomics NFT (non-fungible token) marketplace, and a crypto e-commerce platform was launched for the world's first NFT marketplace to address healthcare verticals, including clinical trials (B2B and B2C). At the moment, Shivom has relationships with worldwide organizations, including universities, institutions, governments, and research organizations.

8.4.5.6 Zenome.io

Zenome [35] is a decentralized blockchain-based genomic database. This platform provides access for the user to manage their valuable genomic data and protect their privacy; it also helps to generate revenue by selling their genomic value to the medical industry in a protected way. It will create equitable opportunities for drug development, and scientific and medical advancement. It is used for storing distributed databases with big genomic data and provides privacy and protected access to millions of human genomes worldwide. It aims to encourage genomic research in emerging markets and demonopolize genetic data in rich nations.

A node is a Zenome participant who contributes computer resources (storage and CPU time) to the distributed storage and processing of genetic data in exchange for

ZNA tokens, or someone willing to contribute genetic information to the Zenome system in order to benefit from selling personal data or using genetics-based services. Zenome software creates a personal account using the GUI (graphical user interface) to receive, transfer, and spend tokens for paid genetic services. Handling NGS genomic data (and most other genomics data) generally involves two steps: preliminary data processing; and dedicated examination of genetic sequences for personal suggestions or research.

The platform is arranged as high-level interactions and a basic system layer (critical infrastructure).

Genomic data are disseminated in a DHT Kademlia network. Participants offering resources for the network are paid in ZNA tokens. They must verify to the network that they are indeed storing the data in order to get paid.

8.4.6 Blockchain-Based Clinical Data Sharing FHIRChain

The healthcare sector has struggled with security and scalability issues for decades. Attempts to merge clinical data have been costly, time-consuming and resulted in data being placed inside separate silos. Consequently, information is not efficiently exchanged and decision-making about patient care is hampered.

Zhang et al. proposed a blockchain-based clinical data sharing architecture system called Fast Healthcare Interoperability Resources (FHIR), suitable for a wide range of healthcare applications [36]. A case study of cancer treatment via telemedicine demonstrated how it may be further modified to enable collaborative clinical decision-making.

FHIR was developed in four major phases. First, the requirements of the Office of the National Coordinator for Health Information Technology (ONC) and the implications of these criteria for blockchain-based systems were thoroughly investigated. Second, in order to meet ONC requirements, FHIR encapsulated HL7 standard-based blockchain architecture for shared clinical data. The third phase consisted of a trial of collaborative decisions on remote cancer treatment, and an FHIR-based decentralized software program was developed that uses digital health IDs to authenticate participation. In the final phase, it may be customized for a specific case study, based on prior learning experiences (Figure 8.8).

Blockchain components connecting medical practitioners and the health care industries are the main components of the architectural perspective diagram. FHIR standardizes the various database symbols and provides a unified, standard structure for sharing of clinical data silos.

Data repositories held in siloed databases are connected to the blockchain via secure database connectors that use secure access tokens to provide authorization to source data references. A smart contract (linked documents) serves as a method of decentralized access and traceability for the secure tokens. The smart contract stores the timestamped log of transactions of all tokens exchanged and consumed, which are immutable. The logs provide information about who has accessed what, how, and when. To ensure data integrity, the FHIRchain should only allow participation by qualified physicians and healthcare organizations on a membership register.

The DApp created using FHIRChain technology [5] illustrates the extent to which blockchain technology may aid in exchanging healthcare data while simultaneously safeguarding the integrity of the original data. FHIRChain may be expanded to include additional players in the healthcare sector, such as insurance firms, and it can make it simpler for individuals to manage their own medical data.

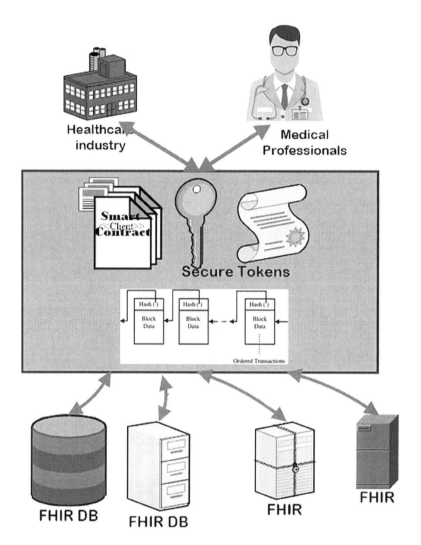

FIGURE 8.8
Architectural view of FHIR chain.

8.5 Conclusion

Blockchain technology combined with IoT technologies is revolutionizing aspects of the healthcare sector, by making data more transparent and accessible to a broader range of users. Technical advancements such as cloud computing and fog computing allow us to take technology to the next level. The lack of trust, security and programming complexity are the two most significant weaknesses of current implementation schemes. These can be overcome by establishing international standards in this area, for which there is a high demand. Blockchain technology will increase the convenience and security of the healthcare sectorof the future.

Notes

1 https://www.hhs.gov/
2 https://github.com/curoverse/arvados
3 https://github.com/exonum

References

1. Abbas, K., Afaq, M., Ahmed Khan, T., & Song, W. C. (2020). A blockchain and machine learning-based drug supply chain management and recommendation system for smart pharmaceutical industry. *Electronics*, 9(5), 852.

2. de Diego, S., Gonçalves, C., Lage, O., Mansell, J., Kontoulis, M., Moustakidis, S., & Liapis, A. (2019, October). Blockchain-based threat registry platform. In *2019 IEEE 10th Annual Information Technology, Electronics and Mobile Communication Conference (IEMCON)* (pp. 0892–0898). IEEE.

3. Brogan, J., Baskaran, I., & Ramachandran, N. (2018). Authenticating health activity data using distributed ledger technologies. *Computational and Structural Biotechnology Journal*, 16, 257–266.

4. Nandi, M. L., Nandi, S., Moya, H., & Kaynak, H. (2020). Blockchain technology-enabled supply chain systems and supply chain performance: A resource-based view. *Supply Chain Management: An International Journal*, 25(6), 841–862.

5. Zhang, P., White, J., Schmidt, D. C., Lenz, G., & Rosenbloom, S. T. (2018). FHIRChain: Applying blockchain to securely and scalably share clinical data. *Computational and Structural Biotechnology Journal*, 16, 267–278.

6. Ackerman, A., Chang, A., Diakun-Thibault, N., Forni, L., Landa, F., Mayo, J., & van Riezen, R. (2016). Blockchain and health IT: Algorithms, privacy and data. *Project Pharm Orchard of MIT's Experimental Learning "MIT FinTech: Future Commerce", White Paper August*, 1–11.

7. Biswas, S. (2020). Measuring performance of healthcare supply chains in India: A comparative analysis of multi-criteria decision making methods. *Decision Making: Applications in Management and Engineering*, 3(2), 162–189.

8. Agbo, C. C., & Mahmoud, Q. H. (2020). Blockchain in healthcare: Opportunities, challenges, and possible solutions. *International Journal of Healthcare Information Systems and Informatics (IJHISI)*,15(3), 82–97.

9. Alzoubi, Y. I., Osmanaj, V. H., Jaradat, A., & Al-Ahmad, A. (2021). Fog computing security and privacy for the internet of thing applications: State-of-the-art. *Security and Privacy*, 4(2), e145.

10. Badhotiya, G. K., Sharma, V. P., Prakash, S., Kalluri, V., & Singh, R. (2021). Investigation and assessment of blockchain technology adoption in the pharmaceutical supply chain. *Materials Today: Proceedings*, 46(20), 10776–10780.

11. Lin, Zhen, Owen, B Art, & Altman, B. Russ. (2004). Genetics. Genomic research and human subject privacy. *Science*, 305, 183183.

12. Mamoshina, P., Ojomoko, L., Yanovich, Y., Ostrovski, A., Botezatu, A., Prikhodko, P., & Zhavoronkov, A. (2018). Converging blockchain and next-generation artificial intelligence technologies to decentralize and accelerate biomedical research and healthcare. *Oncotarget*, 9(5), 5665.

13. Omar, I. A., Jayaraman, R., Debe, M. S., Salah, K., Yaqoob, I., & Omar, M. (2021). Automating procurement contracts in the healthcare supply chain using blockchain smart contracts. *IEEE Access*, 9, 37397–37409.

14. Pattengale, N. D., & Hudson, C. M. (2020). Decentralized genomics audit logging via permissioned blockchain ledgering. *BMC Medical Genomics*, 13(7), 1–9.

15. Sakhi John. (2018) Electronic medical record for deliverance of effective healthcare delivery: Ethical issues and challenges of digitalization in clinical information and Electronic Medical Records (EMR) management. *IOSR Journal of Business and Management (IOSR-JBM)*, 20.3, 01–06.

16. Reda, M., Kanga, D. B., Fatima, T., & Azouazi, M. (2020). Blockchain in health supply chain management: State of art challenges and opportunities. *Procedia Computer Science*, *175*, 706–709.

17. Dunning, T. (2018). Medicines and older people with diabetes: Beliefs, benefits and risks. In *The Art and Science of Personalising Care with Older People with Diabetes*, Springer Charm, 99–120.

18. Kumar, R., & Tripathi, R. (2019, January). Traceability of counterfeit medicine supply chain through Blockchain. In *2019 11th International Conference on Communication Systems & Networks (COMSNETS)* (pp. 568–570). IEEE.

19. Sahoo, M., Singhar, S. S., & Sahoo, S. S. (2020). A blockchain based model to eliminate drug counterfeiting. *Machine Learning and Information Processing;* Springer: Berlin, Germany, 213–222.

20. Bocek, T., Rodrigues, B. B., Strasser, T., & Stiller, B. (2017, May). Blockchains everywhere-a usecase of blockchains in the pharma supply-chain. In *2017 IFIP/IEEE Symposium on Integrated Network and Service Management (IM)* (pp. 772–777). IEEE.

21. Shabani, M. (2019). Blockchain-based platforms for genomic data sharing: A de-centralized approach in response to the governance problems?. *Journal of the American Medical Informatics Association*, *26*(1), 76–80.

22. Uddin, M. (2021). Blockchain Medledger: Hyperledger fabric enabled drug traceability system for counterfeit drugs in pharmaceutical industry. *International Journal of Pharmaceutics*, *597*, 120235.

23. Upadhyay, A., Mukhuty, S., Kumar, V., & Kazancoglu, Y. (2021). Blockchain technology and the circular economy: Implications for sustainability and social responsibility. *Journal of Cleaner Production*, *293*(1), 126130–126140.

24. Velmovitsky, P. E., Bublitz, F. M., Fadrique, L. X., & Morita, P. P. (2021). Blockchain applications in health care and public health: Increased transparency. *JMIR Medical Informatics*, *9*(6), e20713.

25. Akiri. 2021. "The world's first network-as-a-service optimized for healthcare." n.d. https://akiri.com/ (accessed Aug. 17, 2021).

26. "Transform your data into smart data to power AI in a connected, dynamic world," *Burst IQ*, August17, 2021. https://www.burstiq.com/ (accessed Aug. 17, 2021).

27. "Factom Blockchain," *Factom Blockchain*. n.d. https://www.factom.com/ (accessed Aug. 17, 2021).

28. TIMELINES et al., 2021, "Procredex." n.d. https://procredex.com (accessed Aug. 17, 2021).

29. "Medicalchain." n.d.https://medicalchain.com/en/ (accessed Aug. 17, 2021).

30. Grishin, D., Obbad, K., Estep, P., Quinn, K., Zaranek, S. W., Zaranek, A. W., & Church, G. (2018). Accelerating genomic data generation and facilitating genomic data access using decentralization, privacy-preserving technologies and equitable compensation. *Blockchain in Healthcare Today*, 1: 1–23.

31. Jin, X. L., Zhang, M., Zhou, Z., & Yu, X. (2019). Application of a blockchain platform to manage and secure personal genomic data: A case study of LifeCODE.ai in China. *Journal of Medical Internet Research*, *21*(9), e13587.

32. Guðbjartsson, H., Georgsson, G. F., Guðjónsson, S. A., Valdimarsson, R. Þ., Sigurðsson, J. H., Stefánsson, S. K., ... & Stefánsson, K. (2016). GORpipe: A query tool for working with sequence data based on a Genomic Ordered Relational (GOR) architecture. *Bioinformatics*, *32*(20), 3081–3088.

33. "ArvadosUser Guide", n.d. https://doc.arvados.org/v1.3/user/index.html (accessed Aug. 17, 2021).

34. "Exonum Documentation," *ExonumDocumentation*. n.d. https://exonum.com/doc/version/latest/ (accessed August 17, 2021).

35. "A First Decentralized Internet of Genomic Data and Services", n.d. https://zenome.io/ (accessed Aug. 17, 2021).

36. Zheng, X., Sun, S., Mukkamala, R. R., Vatrapu, R., & Ordieres-Meré, J. (2019). Accelerating health data sharing: A solution based on the internet of things and distributed ledger technologies. *Journal of Medical Internet Research*, *21*(6), e13583.

9

Blockchain in Healthcare, Supply-Chain Management, and Government Policies

Satpal Singh Kushwaha, Amit Bairwa, Sandeep Joshi, Sandeep Chaurasia, and Prashant Hemrajani

Manipal University, Jaipur, India

Kailash Kumar

College of Computing and Informatics, Saudi Electronic University, Riyadh, Saudi Arabia

CONTENTS

9.1 Introduction

Blockchain eliminates the need for a trusted third party or central authority in transactional operations because of the consensus algorithms based on cryptographic algorithms [1]. Apart from this, blockchain also provides decentralization, transparency, and immutability, making it a useful technology in a wide variety of domains like education, healthcare, government policies, the internet of things, and supply-chain management. Figure 9.1 shows the basic structure of blockchain [2].

There are four types of blockchain:

- Public blockchains are open for all to join or participate in. Famous examples of public blockchain are Bitcoin and Ethereum [3].

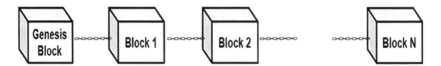

FIGURE 9.1
Blockchain structure.

- Private blockchains have restrictions on participation and mining. A famous example of private blockchain is IBM's Hyperledger [4].
- Consortium blockchains are authorized blockchains with relaxations and restrictions in different categories [5].
- Hybrid blockchains are combinations of private and public blockchain with a publicly available ledger, but modifications to the ledger can be authorized [6].

To satisfy the widely varying needs of individual organizations and clients, numerous blockchain networks are created, each containing a unique arrangement of highlights. However, the structure is the same for all [7]. Bitcoin, the first and the best permissionless blockchain framework, is used in the following example of the critical segments of an ordinary information stream in a blockchain network:

Block: An information structure used to gather a set of exchanges. Information storage is protected by enhanced guarantees. This segment is fundamental to all blockchain networks [8].

Wallet: A safe vault for a client to store their private and public key pair. It is associated with the Bitcoin network so a client can get and send computerized cash (Bitcoins) and screen their equilibrium [9].

Node: A customer who participates in conditional exercises on the blockchain network. First and foremost, a node claims a total and enduring duplicate of every recorded exchange, which applies to every node in a blockchain network. Second, a node adds to the organization by communicating exchanges and empowering diggers to approve and make blocks [10].

A Miner: A digger, an extraordinary client in the Bitcoin organization, gathers and approves all communicated exchanges and makes new squares. It contends with different diggers in the organization to tackle a numerical riddle, generally known as a proof-of-work issue. The first to win the riddle adds another square to the chain and gains a specific measure of remuneration, like a few Bitcoins. At the point when a square is added, all nodes synchronize their neighbouring duplicate, guaranteeing that their record is unique. Digging methodology is utilized for approval in numerous permissionless blockchains, while approval is executed by hubs heavily influenced by an agreement in most permissioned blockchains [11].

Consensus algorithm: An agreement between nodes in a blockchain network that submits conditional data. This is the most basic segment of blockchain. A blockchain network is refreshed by means of an agreement convention which guarantees that exchanges and squares are requested effectively. Thus the integrity and consistency of the disseminated record is ensured, and, finally, trust between partners (nodes) is enhanced [12].

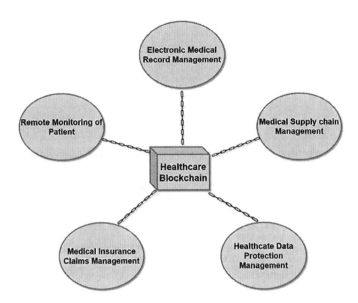

FIGURE 9.2
Blockchain in healthcare.

9.2 Blockchain in Healthcare

Blockchain can be potentially utilized in the field of healthcare, where it can form a stand-alone framework for storing continually updated health records for rapid and secure access by approved clients. Miscommunication between various medical professionals dealing with the same patient can be avoided, enabling more rapid diagnosis and customized care. Figure 9.2 shows potential uses of blockchain in healthcare [13].

9.2.1 Electronic Medical Record Management

Blockchain provides a standalone exchange layer where information can be submitted, shared and stored through a single secure framework, using private encoded connections to store X-ray or other images [14]. The system can be kept consistently available through agreements and uniform approval conventions.

9.2.2 Remote Patient Monitoring

Portable health applications are becoming more significant , which encourages innovation. Electronic medical records (EMRs) of remote patient monitoring can be kept securely in a blockchain network. The information can rapidly be accessed by the clinical workforce and the patient [15]. This area is vulnerable to malware, particularly root access which can allow the malware programmer to use the patient's private key.

9.2.3 Medical Supply Chain Management

Blockchain can conveniently secure, and record the path of, drug supplies. It can even document associated costs and fossil fuel emissions [16].

9.2.4 Medical Insurance Claims Management

Blockchain is particularly suitable for the recording of clinical incidents as they happen, without the potential for fraudulent alteration of the information after the event [17].

9.2.4.1 Case Study

Blockchain can be used for the efficient management of medical insurance claims. Figure 9.3 shows a case study of this.

In a medical insurance claim each stakeholder has their own role as follows:

Patient: The patient uploads information including their medical history, medical insurance details, proof of identity and bank account details. The patient requests medication from a hospital by providing their insurance company details.

Hospital: The hospital processes medication requests from the patient by fetching their medical history and insurance details. After verifying the insurance details with the insurance company, the hospital can initiate medical consultation. The hospital then instructs a pharmacy to supply the medicines to the patient, and invoices the insurance company for settlement of the medical bill.

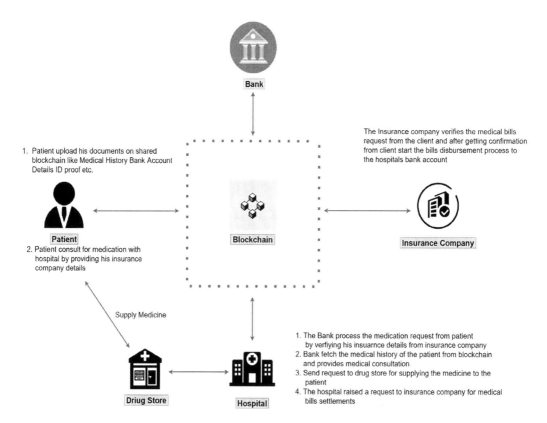

FIGURE 9.3
Blockchain use case in medical insurance claims management.

The Insurance Company: After receiving an invoice for medical bill settlement, the insurance company verifies the bill with the patient and disburses funds to the hospital's bank account.

9.2.5 Healthcare Data Protection Management

Between 2009 and 2017 more than 176 million pieces of information were added to health care records. Blockchain can ensure the security of this health data. Every individual has both a public identifier or key and a private key, which can be opened uniquely as required for the relevant period. Hacking to gain access to the records would be restricted by the need to assault every client independently. Subsequently, blockchains can give a permanent audit trail of health data [18].

9.3 Blockchain in Supply Chain Management

Supply chains improve effectiveness, responsiveness and output in present-day industries. In recent years organizations have expanded, the geographic range involved in production has increased, and item portfolios have broadened [19]. The supply chain has developed from a simple organization of producers and providers to a complex matrix of different items travelling through various gatherings and requiring the participation of many partners. Because e-commerce has developed so rapidly, interest in item traceability and source-to-store tracking has never been higher. Shortcomings of information concerning the current supply chain have significantly affected the activities of retailers and makers [20]. Figure 9.4 shows the use of blockchain in the supply chain.

Previously gaps in the data gathered by factories and retailers made it difficult to follow the history of an item. Blockchain offers an improved supply chain that incorporates the start-to-finish stream, encompassing both the physical and associated information streams of raw materials, finished items, data, and cash [21]. It can play a central role in the functioning of organizations. The supply chain administers, or is associated with, sourcing, acquisition, manufacture, transport and coordination. It thus influences speed-to-showcase, the final price of an item, administration insight, and capital requirements within organizations [22]. The supply chain coordinates many divided and often topographically discrete cycles into a firm framework to convey value to the client [23].

At present supply chains present the following problems and challenges:

- **Traceability:** Traceability has become vital for supply chains, especially for client care, and business organization and planning. It is hard to establish a coherent traceability framework in an interrelated organization, particularly where trust between participants are restricted. Traceability is a crucial aspect of supply chain management that requires appropriate mechanisms [24].
- **Lack of trust between stakeholders:** Trust is an essential factor in the supply chain. A robust network should be built on strong trust. Doubt among members is the most important barrier to the further development of production network organizations [25]. Most organization partners depend on outside agents to act as trust specialists and to validate transactions, which greatly increases cost and reduces productivity.

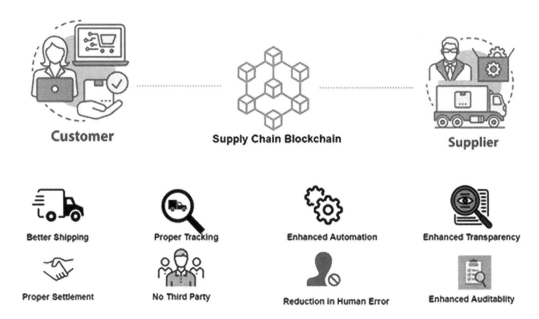

FIGURE 9.4
Blockchain in supply chain.

- **Lack of transparency:** A straightforward supply chain network enhances trust among partners and ensures the quality of items and related information. Discrete databases in current inventory chain networks offer minimal transparency, and much of their useful information is lost when objects and data are transferred between partners. There are also problems with contradictory data, reliance on paper credentials, and deficient interoperability [26]. Such fundamental problems remain despite long periods of research investment.

- **Problems in data sharing:** In current supply chain networks, information is divided between partner organizations using paper documents. Frequently substantial records, like bills of lading, letters of recognition, solicitations, safety cases, and separate declarations, must accompany their associated merchandise everywhere [27]. Constrained by this superseded and extravagant data distribution technique, ships and aircraft are frequently delayed when paperwork doesn't coincide with product movement.

- **Compliance with national standards:** Companies must meet increasingly rigorous managerial principles to provide reliable products and services to customers. The U.S. Food and Drug Administration and the Federal Trade Commission set standards for food handling and traceability of food movement throughout the supply chain [15]. With existing manufacturing system measures, it is difficult for a group of allied producers to provide an information base that conforms to the standards.

Blockchain is seen as an innovation with the inherent capacity to further develop supply chain productivity. Variants of blockchain can considerably ease or even eliminate the supply chain issues mentioned above. Blockchain can improve supply-chain viability, effectiveness and simplicity, and could reduce conditional time and cost [28]. With the introduction of blockchain, traceability inside the store network is significantly improved, creating a completely auditable path of everything moving through the organization. When combined with Internet-of-Things devices, such as RFID tags, a blockchain-empowered production network can continuously gather 'thing-level' information concerning extremely large numbers of items [29]. Such data includes timestamps and assortment areas to create an audit trail that is finished, precise, and simple to access, from the item's starting point to client delivery [30].

Because blockchain information is permanent and includes digital signatures that confirm data proprietorship, information stored in the blockchain offers a protected and full history of anything in the whole store network. In the case of a faulty item, improved detectability allows the root cause of the issue to be recognized more rapidly, which reduces both the cost of item review and disturbance between partners. Enhanced traceability gives stakeholders and clients more trust in an item's validity and quality.

- **Enhanced transparency:** Blockchain gives the executives in the inventory network the ability to see who is performing what activities, at what time, and where. This data is stored in dispersed records that can easily be accessed by included and confirmed partners. Through the coordination of physical and computerized streams across the store network, the efficiency of various interactions will improve. A blockchain-empowered inventory network, with its straightforward and complete item stock stream, assists organizations with improving estimates and choices [31]. It is also a useful asset for combating extortion and fraud.

- **Enhanced efficiency:** One important reason to introduce blockchain is to replace outdated, paper-based processes. The incorruptible data records provide real-time neighbourhood clones for all parties within the company. All transactions are immediately echoed into every neighbourhood copy [31]. Blockchain enhances trading reliability and speed by diminishing individual mistakes and 'eliminating the middleman'.

- **Enhanced security:** Blockchain records are extremely resistant to hacking assaults like those that have compromised the concentrated data sets of intermediaries such as banks. They are organized in such a way that an attempted hack into a particular block will alter all previous blocks. Blockchain is therefore a safer way to log business processes and transactions.

- **Enhanced trust:** Transactions within a blockchain-based network are created and recorded by distributed communication that can be trusted by all participants. Time, location, and specific information relating to each activity on an item in the inventory network are all recorded. All information is synchronized to all partners in real-time, which improves trust among partners inside the production network organization.

- **Ease of Compliance:** A blockchain-empowered production network records all exchanges with exact details, including timestamps and location. These precise, sealed records can assure the trustworthiness of business information and can be readily accessed to comply with guidelines and consistency [32].

9.4 Blockchain in Government Policies

Governments are evaluating the likely uses of blockchain technology in the public area. Appropriate government records may form another data framework supporting the exchange of data between government departments, citizens, and organizations. Particular groups of use-cases that employ decentralized data structures have been identified for local populations. Blockchain technology has been specifically employed within the different taxpayer-driven organizations and services. Examples include the maintenance of resident records, central banks and monetary exchanges, electronic voting in support of democracy, administrative oversight of business sectors, combating tax fraud/avoidance and the allocation of public funds, awards, social exchanges and pensions [33].

Digital government is a concept derived from policy management science, a replacement of the e-government worldview. The previous model encompassed the digitization of policy management. Fully digital government refers to the formation of new open administrations and administration conveyance models that influence computerized advances and legislative and resident data resources. The new concept centres around the arrangement of client-driven Spry (a digital vendor) and creative public administrations. Blockchain is one of the foremost innovations to be considered for the new worldview of administrative strategy making and administration conveyance. Figure 9.5 shows blockchain use cases in government policies.

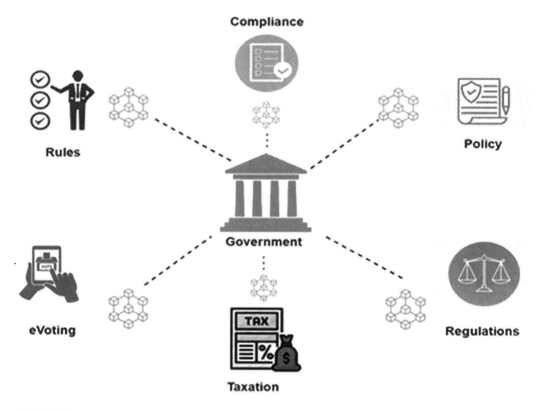

FIGURE 9.5
Blockchain in government.

The primary advantages of applying blockchain in government are proposed as:

- Reduced cost, time, and complexity in data exchange between legislative, public and private bodies, which in turn upgrades the authoritative capacity of governments.
- Automated smart contracts ultimately reduce the requirement for records to be circulated for approval.
- Increased simplicity, auditability, and reliability of data in administrative records that support citizens.
- Increased trust of citizens and organizations in administrative cycles and record keeping, driven by the use of calculations that are currently not under the sole control of the government.

With regard to advanced government, blockchain can enable direct connections between open foundations, citizens, and monetary specialists. Essentially, this means open administration in data enrollment and trade measures. Blockchain is a combination of a few existing innovations that provide decentralized data structures. The decentralization of blockchains is the key to reshaping the way governments interface with citizens and with one another (Atzori, 2015). Blockchain could greatly reduce the managerial overheads that legislatures currently impose in the public arena.

Governments alone perhaps don't need to provide data collection and exchange to plan the financing of social support, as this could be provided by the blockchain convention. They should simply maintain administrative oversight of exchanges occurring in this framework. Blockchain might be used as a data framework for trading data between open organizations. For instance, exchange of data concerning criminal convictions, awards, scholastic degrees or duties could employ blockchain (Davidson, De Filippi, and Potts, 2016). Dispersed holding of data, rather than single-point records, is believed to offer specialized and monetary benefits. Simplicity, unwavering quality, and the possibility for more elaborate execution are particularly significant when applications require information obtained from different places, associations, or nations. Unexpectedly, the shared data in blockchain frameworks reduces vulnerabilities in the robustness the organization, as it eliminates one element of control. For instance, at present banks hold unique power in the framework, whilst in a blockchain-based framework power in the network is shared among all the members.

Decentralization is, to some extent, challenging as it conflicts with the present institutional designs of governments, partnerships, and commercial centres. Governments in particular ought to consider the administrative and hierarchical benefits of blockchain processes, given their major contrast with conventional data frameworks. To completely exploit the benefits of blockchain in the public area, responsibilities and processes need to be re-designed around the innovation and not the other way round. The following could be use cases for blockchain technology:

- Management of citizen identity
- Tax returns
- Education management
- Electronic voting
- Governing compliance

Blockchain likewise offers benefits to citizens. In particular, citizens can gain financial benefits and efficiency gains from administrative processes that entail legal agreement or authentication, such as property conveyancing and land titles. Administrations that draw on the decentralized nature of blockchain, like identity or voting, exert an overall influence, expanding the proprietorship and control of citizens. Given these advantages and difficulties, the introduction of blockchain disturb the situation in the public area. Blockchain can improve effectiveness by crossing siloes, smoothing levels and introducing new models for government. The structure of blockchain can diminish functional risk and costs, improve consistency and trust in government establishments.

9.5 Conclusion

Blockchain technology offers great potential to improve the present production network. It provides decentralized storage to share any kind of transactional data in an unalterable and extremely robust form. We believe it has unparalleled potential for store networks, as it promises to deliver an effective structure for organizations to share information quickly and securely across a variety of store network areas and cycles. This invention allows businesses to create a more flexible and capable store network, and to thoroughly address external and internal challenges. Despite the promise of blockchain technology, there is a need for more exploration to fully assess this innovation. Ongoing work is removing impediments in adaptability, security, and protection to gain trust in the application of this innovation to medical care. Blockchain has not yet broken ground, nor even have its potential problems been investigated, for the public area. We have not seen the generation of new plans of action, the development of new administrative paradigms nor any direct disintermediation of public organizations associated with the arrangement of administrative capacities. Genuinely innovative administrations which provide democratic decentralization of city administration without direct government intervention are absent from the current scene.

References

1. S. Khezr, M. Moniruzzaman, A. Yassine, and R. Benlamri, "Blockchain technology in healthcare: A comprehensive review and directions for future research," *Appl. Sci.*, vol. 9, no. 9, pp. 1–28, 2019, doi: 10.3390/app9091736
2. S. Alharthi, P. R. C. Cerotti, and S. Maleki Far "An Exploration of the Role of Blockchain in the Sustainability and Effectiveness of the Pharmaceutical Supply Chain," *J. Supply Chain Cust. Relatsh. Manag.*, vol. 2020, pp. 1–29, 2020, doi: 10.5171/2020.562376
3. S. Saberi, M. Kouhizadeh, J. Sarkis, and L. Shen, "Blockchain technology and its relationships to sustainable supply chain management," *Int. J. Prod. Res.*, vol. 57, no. 7, pp. 2117–2135, 2019, doi: 10.1080/00207543.2018.1533261
4. A. Kumar, R. Krishnamurthi, A. Nayyar, K. Sharma, V. Grover, and E. Hossain, "A Novel Smart Healthcare Design, Simulation, and Implementation Using Healthcare 4.0 Processes," *IEEE Access*, vol. 8, pp. 118433–118471, 2020, doi: 10.1109/ACCESS.2020.3004790

5. T. Le Zhu and T. H. Chen, "A Patient-Centric Key Management Protocol for Healthcare Information System based on Blockchain," *2021 IEEE Conf. Dependable Secur. Comput. DSC 2021*, vol. 1, pp. 1–5, 2021, doi: 10.1109/DSC49826.2021.9346259

6. S. Chakraborty, S. Aich, and H. C. Kim, "A Secure Healthcare System Design Framework using Blockchain Technology," *Int. Conf. Adv. Commun. Technol. ICACT*, vol. 2019-February, pp. 260–264, 2019, doi: 10.23919/ICACT.2019.8701983

7. M. Rhies Khan and A. Manzoor, "Application and Impact of New Technologies in the Supply Chain Management During COVID-19 Pandemic: A Systematic Literature Review," *Int. J. Econ. Bus. Adm.*, vol. IX, no. 2, pp. 277–292, 2021, doi: 10.35808/ijeba/703

8. W. Alshahrani and R. Alshahrani, "Assessment of Blockchain technology application in the improvement of pharmaceutical industry," *2021 Int. Conf. Women Data Sci. Taif Univ. WiDSTaif 2021*, pp. 14–18, 2021, doi: 10.1109/WIDSTAIF52235.2021.9430210

9. I. A. Omar, R. Jayaraman, M. S. Debe, K. Salah, I. Yaqoob, and M. Omar, "Automating Procurement Contracts in the Healthcare Supply Chain Using Blockchain Smart Contracts," *IEEE Access*, vol. 9, pp. 37397–37409, 2021, doi: 10.1109/ACCESS.2021.3062471

10. W. Bodeis and G. P. Corser, "Blockchain Adoption, Implementation and Integration in Healthcare Application Systems," *Conf. Proc. - IEEE SOUTHEASTCON*, vol. 2021-March, pp. 5–7, 2021, doi: 10.1109/SoutheastCon45413.2021.9401885

11. K. Khujamatov, E. Reypnazarov, N. Akhmedov, and D. Khasanov, "Blockchain for 5G Healthcare Architecture," *2020 Int. Conf. Inf. Sci. Commun. Technol. ICISCT 2020*, 2020, doi: 10.1109/ICISCT50599.2020.9351398

12. P. P. Ray, D. Dash, K. Salah, and N. Kumar, "Blockchain for IoT-Based Healthcare: Background, Consensus, Platforms, and Use Cases," *IEEE Syst. J.*, vol. 15, no. 1, pp. 85–94, 2021, doi: 10.1109/JSYST.2020.2963840

13. T. K. Dasaklis, F. Casino, and C. Patsakis, "Blockchain Meets Smart Health: Towards Next Generation Healthcare Services," *2018 9th Int. Conf. Information, Intell. Syst. Appl. IISA 2018*, pp. 1–8, 2019, doi: 10.1109/IISA.2018.8633601

14. S. Wang et al., "Blockchain-Powered Parallel Healthcare Systems Based on the ACP Approach," *IEEE Trans. Comput. Soc. Syst.*, vol. 5, no. 4, pp. 942–950, 2018, doi: 10.1109/TCSS.2018.2865526

15. A. R. Lee, M. G. Kim, K. J. Won, I. K. Kim, and E. Lee, "Coded Dynamic Consent Framework Using Blockchain for Healthcare Information Exchange," *Proc. - 2020 IEEE Int. Conf. Bioinforma. Biomed. BIBM 2020*, pp. 1047–1050, 2020, doi: 10.1109/BIBM49941.2020.9313330

16. S. Vyas, M. Gupta, and R. Yadav, "Converging Blockchain and Machine Learning for Healthcare," *Proc. - 2019 Amity Int. Conf. Artif. Intell. AICAI 2019*, pp. 709–711, 2019, doi: 10.1109/AICAI.2019.8701230

17. N. Kshetri, "1 Blockchain's Roles in Meeting Key Supply Chain Management Objectives," *Int. J. Inf. Manage.*, vol. 39, no. December 2017, pp. 80–89, 2018, doi: 10.1016/j.ijinfomgt.2017.12.005

18. M. Zarour et al., "Evaluating the Impact of Blockchain Models for Secure and Trustworthy Electronic Healthcare Records," *IEEE Access*, vol. 8, pp. 157959–157973, 2020, doi: 10.1109/ACCESS.2020.3019829

19. A. Yogeshwar and S. Kamalakkannan, "Healthcare Domain in IoT with Blockchain Based Security - A Researcher's Perspectives," *Proc. - 5th Int. Conf. Intell. Comput. Control Syst. ICICCS 2021*, no. Iciccs, pp. 440–448, 2021, doi: 10.1109/ICICCS51141.2021.9432198

20. C. Agbo, Q. Mahmoud, and J. Eklund, "Blockchain Technology in Healthcare: A Systematic Review," *Healthcare*, vol. 7, no. 2, p. 56, 2019, doi: 10.3390/healthcare7020056

21. W. Kersten, T. Blecker, and M. Ringle Christian, Digitalization in Supply Chain Management and Logistics : Smart and Digital Solutions for an Industry 4.0 Environment. Berlin: epubli GmbH; 2017.

22. D. Shakhbulatov, J. Medina, Z. Dong, and R. Rojas-Cessa, "How Blockchain Enhances Supply Chain Management: A Survey," *IEEE Open J. Comput. Soc.*, vol. 1, no. June, pp. 230–249, 2020, doi: 10.1109/ojcs.2020.3025313

23. A. Farouk, A. Alahmadi, S. Ghose, and A. Mashatan, "Blockchain Platform for Industrial Healthcare: Vision and Future Opportunities," *Comput. Commun.*, vol. 154, pp. 223–235, 2020, doi: 10.1016/j.comcom.2020.02.058

24. M. Kassab, J. Defranco, T. Malas, G. Destefanis, and V. V. G. Neto, "Investigating Quality Requirements for Blockchain-Based Healthcare Systems," *Proc. - 2019 IEEE/ACM 2nd Int. Work. Emerg. Trends Softw. Eng. Blockchain, WETSEB 2019*, pp. 52–55, 2019, doi: 10.1109/WETSEB.2019.00014

25. P. Pandey and R. Litoriya, "Promoting Trustless Computation Through Blockchain Technology," *Natl. Acad. Sci. Lett.*, vol. 44, no. 3, pp. 225–231, 2021, doi: 10.1007/s40009-020-00978-0

26. M. Kaur, M. Murtaza, and M. Habbal, "Post study of Blockchain in Smart Health Environment," *CITISIA 2020 - IEEE Conf. Innov. Technol. Intell. Syst. Ind. Appl. Proc.*, 2020, doi: 10.1109/CITISIA50690.2020.9371819

27. P. Ndayizigamiye and S. Dube, "Potential Adoption of Blockchain Technology to Enhance Transparency and Accountability in the Public Healthcare System in South Africa," *Proc. - 2019 Int. Multidiscip. Inf. Technol. Eng. Conf. IMITEC 2019*, pp. 17–21, 2019, doi: 10.1109/IMITEC45504.2019.9015920

28. V. Paliwal, S. Chandra, and S. Sharma, "Blockchain Technology for Sustainable Supply Chain Management: A Systematic Literature Review and a Classification Framework," *Sustain.*, vol. 12, no. 18, pp. 1–39, 2020, doi: 10.3390/su12187638

29. J. Qiu, X. Liang, S. Shetty, and D. Bowden, "Towards Secure and Smart Healthcare in Smart Cities Using Blockchain," *2018 IEEE Int. Smart Cities Conf. ISC2 2018* 2019, doi: 10.1109/ISC2.2018.8656914

30. J. Xu, N. Abdelkafi, and M. Pero, "On the Impact of Blockchain Technology on Business Models and Supply Chain Management," *Proc. Summer Sch. Fr. Turco*, pp. 1–7, 2020.

31. G. Mirabelli and V. Solina, "Blockchain and Agricultural Supply Chains Traceability: Research Trends and Future Challenges," *Procedia Manuf.*, vol. 42, no. 2019, pp. 414–421, 2020, doi: 10.1016/j.promfg.2020.02.054

32. E. A. Maroun and J. Daniel, "Opportunities for Use of Blockchain Technology in Supply Chains: Australian Manufacturer Case Study," *Proc. Int. Conf. Ind. Eng. Oper. Manag.*, vol. 2019, no. MAR, pp. 1603–1613, 2019.

33. C. Shen and F. Pena-Mora, "Blockchain for Cities - A Systematic Literature Review," *IEEE Access*, vol. 6, pp. 76787–76819, 2018, doi: 10.1109/ACCESS.2018.2880744

10

Electricity and Hardware Resource Consumption in Cryptocurrency Mining

Lokesh Gundaboina and Sumit Badotra
Lovely Professional University, Punjab, India

Gaurav Malik
Ceridian, Toronto, Ontario, Canada

Vishal Jain
Sharda University, Greater Noida, U.P., India

CONTENTS

10.1 Introduction

The use of cryptocurrency has been growing in recent years, leading to an expansion of interest in cryptocurrency mining. Cryptocurrency mining is a technique that validates transactions for various sorts of cryptocurrencies and adds them to a digital blockchain ledger. It is also known as crypto-coin mining, altcoin mining, or bitcoin mining [1]. Cryptocurrency miners compete with one another to unravel complex math problems using high-performance computer systems and a variety of algorithms [2]. In addition to transaction processing charges in exchange for their services, miners obtain newly generated cryptocurrencies [3]. Not all cryptocurrencies are mineable; bitcoin is the most established and best-known example of a mineable cryptocurrency.

DOI: 10.1201/9781003240310-10

The massive systems used by miners to mine blocks and validate transactions consume large amounts of electricity. The energy consumption of bitcoin miners is relatively easy to estimate by examining its hash rate i.e., the total mixed computational power used to mine bitcoin and process transactions [4]. However, the carbon emissions of the hardware are much more difficult to determine. Accurate estimates of energy production by geolocation, from which an energy blend may be inferred, have been produced by the Cambridge Center for Alternative Finance (CCAF), which has worked with essential mining pools to prepare an anonymized dataset of miner locations [5].

Another key aspect in which Bitcoin's energy consumption differs from that of other industries is that Bitcoin can be minedanywhere [6]. While most energy used globally has to be generated exceptionally close to the end user, Bitcoin has no such restriction, enabling miners to apply electricity sources that might be unreachable for optimum use [7].

10.1.1 Bitcoin

Bitcoin is a digital currency that is backed by a cryptology-protected circulation book and is the most important and well-known blockchain application. The computational validation process known as mining requires specific hardware that depends on a decentralized system of transactions verified by the nodes of the crypto network and recorded in a distributed ledger called the blockchain (Satoshi, 2008). This occurs in over-the-counter [OTC] marketplaces or as an intermediate medium of exchange for various cryptocurrencies, products, or resources [8].

Bitcoin mining was originally the most effective way for computer specialists to accumulate price BTC on the blockchain [4]. BTC became popular when it peaked at US$19,783 and hit a capitalization of US$332 billion for 17.8 million bitcoins in December 2017. This increasing quality has led to diverse sorts of mining rather than sourcing BTC directly from the crypto markets. The massive variation between buying a bitcoin and therefore the annual mining cost has continued to draw in distinct investors from different backgrounds to choose the simplest sorts of cryptocurrency mining possible [9].

10.1.2 How Bitcoin Works

Bitcoin, the most shared place in the world and the famous cryptocurrency, is growing in popularity [10]. Its structure has remained essentially the same since its establishment in 2008, but repeated global market shifts have created a modern cryptocurrency whose name is now worth much more than at its initial introduction. Combined with a network of computer systems that verify transactions, users can exchange hashes as there are a limited number of bitcoins that can ever be generated, which ensures their rarity. Bitcoin's value lies in the fact that its users, when they accept it as payment, can be sure of its genuineness and can use it elsewhere to make other purchases [10].

To feature a block to the blockchain a trademark must be found that links the transactions within the block. Miners seek out a current price that specifics a particular equation using the SHA256 cryptographic hash function [5]. This is often a calculational overrated task, yet a participant of the peer network who discovers an inexpensive price will be rewarded with the choice of assigning recently mined bitcoins to a particular address for selection.

A generated transaction takes place after a payer has sent a currency to a drawee. Mining confirms transactions and adds them to the current public ledger [9]. As soon as a replacement transaction has been administered, the laborer checks whether or not the coin belongs to the remunerator or whether or not the remunerator has tried to pay double. A malicious user will produce multiple associated nodes in an attempt to validate an invalid group action. To prevent this from happening, miners have to solve a resource-intensive task [11]. The resource intensity makes it too expensive for a malicious user to create enough cast identities.

Today, thousands of people transfer and receive bitcoins every couple of minutes in the peer-to-peer digital currency system developed by Satoshi Nakamoto [1]. Miners are able to check the legality and credibility of transactions, take information from the block, with the option of a variable referred to as a present and execute it from the secure hash algorithmic rule. The rule runs throughout the block and converts the initial information into a 256-bit sequence referred to as a hash [4].

Even a small change within the initial data will forcefully alter the subsequent hash. However, miners do not modify the data and information regarding transactions; they modify the desired variable to produce an individual hash. The aim is to find a hash that options a particular quantity of high zero bits [12]. Put simply, bitcoin miners create cash after finding a 32-bit value that, once processed employing a standard hashing operation together with the information from various transactions, creates an all-new hash with a selected range of 60 [11]. Most people regard crypto mining simply as a way of creating new cash. However, cryptocurrency transactions are verified on a blockchain community and included in an allotted ledger [11]. Most importantly, crypto mining prevents the double-spending of digital currency on an allotted network.

Digital systems are simple to operate [13]. Like physical currencies, as soon as one member spends cryptocurrency, the virtual ledger is updated by debiting one account and crediting the other. Bitcoin's allotted ledger enables proven miners to update organization movements on the digital ledger, making them additionally responsible for securing the network from double-spending [14].

Since distributed ledgers lack a centralized authority, the mining technique is critical for verifying transactions [15]. Miners are incentivized to maintain the network by participating in the transaction validation procedure, increasing their chances of obtaining new-minted cash [2]. To ensure that the best crypto miners can mine and validate transactions, so-called proof-of-work (POW) consensus protocols have been established which secure the network from any outside attacks [2].

Crypto mining is just like mining precious metals. Just as miners of precious metals can dig for gold, silver, or diamonds, crypto miners release modern-day cash into circulation [15]. For miners to be rewarded with new coins, they need to set up machines that can solve advanced mathematical equations in the form of science hashes. A hash is a truncated digital signature of a bit of information. Hashes are generated to store information transferred on a public network [15]. Miners compete to zero in on a hash charge generated through a crypto coin transaction; the first miner to crack the code receives a block on the ledger and earns a reward [4]. Each block makes use of a hash function to consult the preceding block, forming an unbroken chain of blocks leading back to the first block [16].

As miners set up superior machines to resolve POW, the difficulty of equations in the community is increasing. At the same time, opposition amongst miners has been rising, resulting in a reduction of the amount of cryptocurrency in circulation [8] (Figure 10.1).

10.1.3 Organization of the Chapter

This chapter is organized in seven sections. A brief literature survey follows this introduction, after which methodology is explained in the third section. The fourth section contains

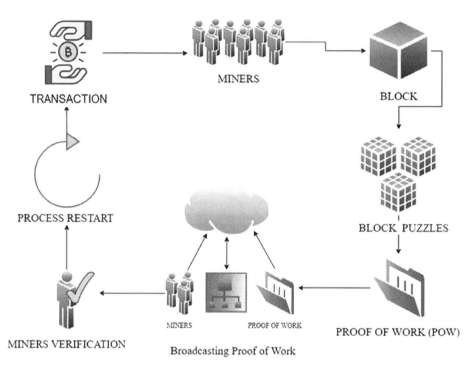

FIGURE 10.1
Working of bitcoin in blockchain.

a discussion, while the fifth enumerates the advantages and disadvantages of mining cryptocurrency. The sixth and seventh sections, respectively, draw conclusions and outline future scope.

10.2 Literature Survey

Bitcoin (BTC) is normally defined as a cryptocurrency and it is now an innovative virtual benefit as a complement to mining technology [17]. Since the publication of the seminal article by Nakamoto Satoshi in 2008 concerning the introduction of BTC as a digital peer-to-peer (P2P) currency system, BTC has become the world's major cryptocurrency for varied cryptocurrencies, products, or services. The growing recognition of bitcoin has led to creation of a variety of mining techniques. The large difference between buying a BTC and the usual mining costs has attracted independent investors from different backgrounds to settle on the simplest options for cryptocurrency mining techniques, from domestic mining to cloud and hosted mining.

Each cryptocurrency system maintains a distributed public ledger referred to as the blockchain [14]. A group action is created when a remunerator causes coins to be transferred to a receiver. Mining validates transactions and makes them available on the current public ledger [18]. With a replacement group action, the miner checks whether or not the coin belongs to the dealer, and whether or not they are attempting to pay cash double.

A malicious person can produce and review a node to validate a transaction. To avoid this, miners are obliged to perform a resource-intensive task.

The Bitcoin network is secured via means of individuals referred to as miners. Any appliance withinside the Bitcoin grid can operate as a miner [14]. Users have used different kinds of hardware to mine Bitcoin blocks over the years. Mining, FPGA mining, and ASIC mining square measure are examples of hardware widely used for Bitcoin mining [2]. All hardware mining has to deal with minimum profits, and excessive power costs. Cloud mining does not incur additional heating or excessive electricity costs, but it has its own limitations. The miner uses their computing power to solve the puzzle and broadcast it to the grid. A transaction is only regarded as legitimate when it is signed by the sender. The mining is concluded when it is verified by miners who are looking at Bitcoin transactions constantly. It takes an average of 10 minutes to add a replacement transaction block to the blockchain within the network.

The Bitcoin blockchain ledger has an average of 120,000 transactions worth US$75 million per day. According to websites such as Blockchain.info and coinmarketcap.com, there are currently more than 1,500 different kinds of cryptocurrencies within blockchain-based markets with a total capitalization of US$500 billion. ETH and Ripple are in second and third places with 37% in total [18]. The system price was US$185.8 billion, roughly equivalent to the size of New Zealand's GDP, in 2016. In other words, decentralized Bitcoin has established a medium-sized world national economy based chiefly on trust and algorithms [4].

The competition between miners multiplied between August 2009 and August 2017 by a factor of 109 [19]. In the same period, the entire computing power of the Bitcoin network increased from 10 GH / s to 9.3 EH /s, and therefore the number of daily transactions has further multiplied from 50 to 300,000 [12]. This competition encourages miners to accumulate greater computing resources. They use state-of-the-art mining devices specially designed to solve the POW problem [15]. Thousands of these power-hungry mining devices, burning 0.8–2.0 kWh of electricity, are used in the mining process. Although the complexity of mining has multiplied, the computing pace of mining devices has not kept pace with it due to the circumstances of 2009. The increase in mining activities, however, raises the question of whether or not the overall power called for by mining is sustainable if Bitcoin reaches a transaction volume similar to that of conventional financial systems.

As the resource required for running Bitcoin has grown in recent years, its capability has begun to have an impact on climate and human health [19]. In this context, a number of organizations are seeking accurate information, for example, in order to assess the urgency of the matter, to provide coverage in the right locations, and to put mitigation programs in place [15]. We propose a market dynamics methodology to assess techniques for assembling information on Bitcoin energy consumption. For most of the time the bitcoin company has been developing this boom it has let the market pursue choices that do not necessarily involve top-quality tools or locations, leading to a decidedly suboptimal performance by the Bitcoin community. Market participants principally use older-generation devices that are more available and lower cost [16]. General estimation strategies do not make the most of this behavior, however, neither do they make the most of the market conditions, like seasonal and geographic instabilities in power costs. Strategies appear to provide positive estimates throughout growth cycles. We cautiously estimate that the Bitcoin network consumes 87.1 TWh of electricity annually [16].

Bitcoin's hunger for energy has sparked a debate in the academic literature and among the general public over the consumption of cryptocurrencies [19]. Mining requires special hardware and lots of power to confirm ownership and transactions [13].

The speed of exchange between Bitcoin and different currencies changes over time [19, 20]. This, in turn, affects the profitability of bitcoin mining. On the other hand, the probability of finding a legitimate block can decrease as a result of the increasing number of people mining Bitcoin. Easier hardware is needed to understand identical success rates [16]. However, since price is also a limiting issue, the newer hardware ought to have a much better hash rate and consume less power.

On January 3, 2009, Satoshi Nakamoto generated the first block of the blockchain, the so-called Genesis block, by hashing the central process unit (CPU) of his PC [20]. Like him, the first miners mined Bitcoin by playing on the computer code program on their personal computers. This was the first generation of mining hardware, with subsequent eras represented by GPUs, FPGAs, and ASICs [8]. Each generation transmits with a specific class of mining hardware. In the second generation, beginning in September 2010, forums came onto the market. The square measure chiefly supported parallel running processors (GPUs), giving that generation an advantage. Exceptional mining hardware is characterized by hash rate, energy consumption per hash and increased prices. For example, the NVIDIA Quadro NVS 3100M with 16 cores, the GPU generation, features a hash load of 3.6 MH / s and power consumption of 14 W, and the FPGA generation features a hash load of 800 MH / s and power usage of 40 W, while the ASIC generation features a hash charge of 300 GH/s and electrical consumption of 175 W [13].

Bitcoin's website claims that Bitcoin mining is deliberately described as resource intensive so that the number of blocks found by miners every day remains stable over time and a finite financial supply is manufactured [21]. Miners verify Bitcoin transactions and receive Bitcoin as a gift. Although economic considerations have created a profitable business model, the nature of the mining process is deliberately resource intensive and causes unreasonable environmental damage through high electricity consumption and emission rates [20]. Backed by its decentralized network, the peer verification process is itself polluting. To calculate this effectively, the hardware of the machine used to mine Bitcoin consumes large amounts of electricity 24 hours a day, generating immense amounts of heat and emissions in the process. It weighs 8.1 pounds and has a power consumption of 0.098 W/Gh. Specialized mining operations have tanks filled with this type of machinery, which generally choose a cold climate to avoid the need for room ventilation [16].

Bitcoin was used here as an example of blockchain technology pollution [22]. Some service providers are known to have high energy consumption in their services and have changed or evolved their technology to change the efficiency of the data. Netflix, for example, has improved its technology while offering the same services, but because of the improved efficiency of encryption data transfer requires less data. The aim here is to advocate that the various digital currencies and applications of blockchain technology represent low energy-intensive alternatives [19] (Table 10.1).

10.3 Methodology

This study consists of various experiments on the mining potential of a number of popular cryptocurrencies currently on the market. A detailed comparison is conducted of the applied mathematics analysis and the specific forms of hardware used. We assess which hardware specifications offer the best result for mining, and which are the most energy efficient in terms of both hardware resources and electricity consumption (Table 10.2).

TABLE 10.1

The Literature Survey Overview

Title	Year	Contribution
Is bitcoin a waste of resources?	2018	Resources wasted by Bitcoin.
Blockchain, Bitcoin, and Ethereum Technology: A Brief Overview.	2018	Blockchain technology involved in bitcoin and Ethereum.
Decarbonizing Bitcoin: Legal and Policy Options to Reduce Energy Consumption of Blockchain Technologies and Digital Currencies.	2018	Decarbonizing of bitcoin
A survey about the cryptocurrency Bitcoin and its mining.	2019	Bitcoin as a cryptocurrency and how it is mined.
Hardware Overclocking to Improve Ethereum Cryptocurrency Mining Efficiency.	2020	Hardware overclocking methods for increasing mining efficiency.
Factors Affecting Cryptocurrency Prices: Evidence for Bitcoin, Ethereum, Dash, Litcoin, and Monero.	2018	Factors influencing cryptocurrency prices.
Blockchain generation withinside the chemical industry: Machine-to-device strength market.	2017	Blockchain technology.
The Energy Consumption of Blockchain Technology: Beyond Myth. A new examine Cryptocurrencies.	2020	Energy consumption of Blockchain.
A new examine Cryptocurrencies.	2019	Cryptocurrencies and current changes
SHA256 dual hardware architecture with compact message expander for Bitcoin mining.	2020	Double SHA-256 hardware architecture.
A brief survey of Cryptocurrency systems.	2016	Cryptocurrency systems survey
Bitcoin Mining and Its Cost.	2017	Bitcoin mining and how cost-effective it is.
The Technology and Economic Determinants of Cryptocurrency Exchange Rates: The Case of Bitcoin.	2017	Financial determinants of cryptocurrency change rates.
Why do cryptocurrencies use a lot of energy?	2018	Why cryptocurrencies are energy-hungry.
Green mining: a method of referring to software program alternate and configuration to energy consumption.	2015	Methodologies for reducing power consumption while mining.
Security of Cryptocurrencies in blockchain technology: State-of-art, challenges, and prospects.	2020	Security of cryptocurrency with the help of blockchain.

The crypto world relies on fossil fuels. As the price of bitcoin increases, the amount of electricity used by miners to mine coins will increase, with a large number of users attracted to log into the bitcoin network [1]. Research from the University of Cambridge suggests that the top bitcoins consume more than 120 terawatt-hours (TWh) every year [4].

In a recent report from Galaxy Digital, shared by the International Energy Agency (IEA), the annual energy consumption of the bitcoin network was stated to be 113.89 terawatt-hours per annum (TWh/year), while bank structures use 263.72 TWh / year, and gold mining consumes around 240.61 TWh/year. In effect, therefore, normal banking systems or gold mining consume twice as much electricity as bitcoin mining [23] (Figure 10.2).

The continuous cycle of block mining provides people all over the world with an incentive to mine bitcoin. Since mining can provide a solid income stream, people are willing to run energy-intensive machines [24], and as the value of the currency hit new highs, the energy consumption of the bitcoin network grew to historic proportions [22]. The whole bitcoin network is currently consuming tons of electricity from a variety of countries (Figure 10.3).

TABLE 10.2

Top 10 Best Cryptocurrencies to Mine in 2021

Coin Name	Price	Market Cap	Hardware to Mine	Hash Power	Power Consumption
Bitcoin	$33,188.68	$622.81B	Antminer S19	95.0 TH/s	3250W
Ethereum	$2,035.38	$237.55B	RTX3080	98 MH/s	500W
Bitcoin Cash	$484.23	$9.10B	Antminer S7	4.73 TH/s	1293W
Binance Coin	$317.12	$48.60B	Bitmain Antminer S9i	13.49 EH/s	1320W
Litecoin	$133.51	$8.95B	Antminer L3+	280 MH/s	1050W
XRP	$0.6295	$29.11B	Bitmain Antminer S7	4.73 TH/s	1293W
Dogecoin	$0.2065	$26.96B	Avalon6	3.6 TH/s	1050W
USD Coin	$1.00	$26.49B	Bitmain Antminer R4	8.7 TH/s	845W
Polkadot	$15.00	$14.63B	Pangolin Whatsminer M3X	12.5 TH/s	2050W
Uniswap	$19.57	$11.49B	Bitmain Antminer T9	12.5 TH/s	1576W

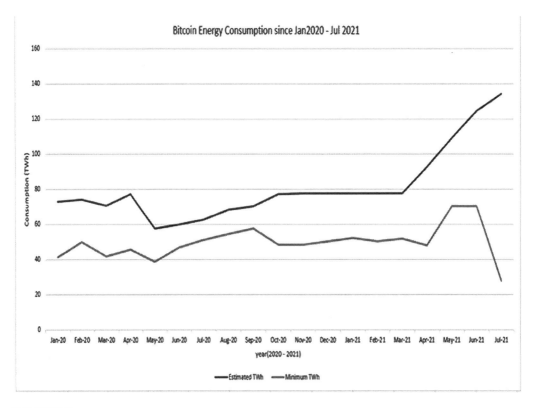

FIGURE 10.2

The energy consumption of bitcoin from January 2020 to July 11, 2021.

The 2017 Global Cryptocurrency Comparative Study states that known facilities with a total consumption of 232 megawatts accounted for about half the full hash rate of bitcoin, with Chinese mining facilities accounting for about half with 111 megawatts [25].

Table 10.3 describes the energy consumption using the estimated factors for each country's network. This service was carried out by Hileman and Rauch [12].

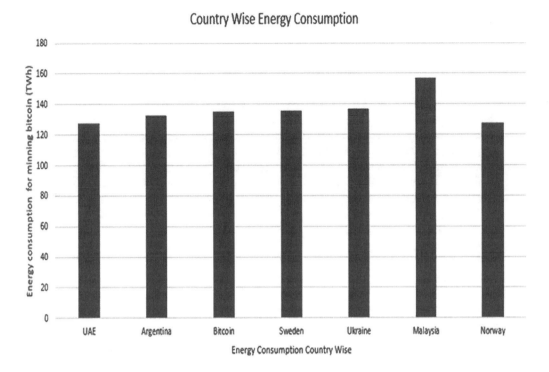

FIGURE 10.3
Country-wise energy consumption of cryptocurrency mining.

TABLE 10.3

Power Consumption and Carbon Intensity

Country	Power Consumption (Megawatts)	Carbon Intensity
China	111	711
Georgia	60	231
U.S	27	486
Canada	18	158
Sweden	10	13
Iceland	5	0
Estonia	2	793
Total/Average	233	475

To reduce the utilization of fossil fuels for mining, miners need to use renewable sources like hydropower [15]. Because of the variations in power generation, bitcoin miners need to virtually manage the use of low-cost hydropower. Crucially, although mining activity takes place in extremely polluting regions of the world, the contribution from renewable energy sources remains low. A survey of miners found that only 39% of their total energy consumption was actually from renewable sources [16].

10.4 Discussion

Mining is a very expensive process as the hardware itself is expensive and there is the added cost of electricity consumption. The most powerful Ant Miner S9 consumes 1.372 kWh and can only produce 14 TH/s, but blockchain's total hash rate is expected to be 10.3 EH/s (1 EH/s = 106 TH/s) [26]. Figure 10.4 illustrates the hash rate share among different countries.

We estimate overall power usage in the bitcoin network using slush pool model parameters. Electricity consumption by network devices such as routers is not included in our analysis. For mining, we assess the most up-to-date and efficient hardware available and use that specific hardware [19]. For every 100% rise in transaction volume, we estimate that the computing power of mining machines will increase by 25% and energy usage will decrease by 10%. We estimate overall electricity consumption as bitcoin's network increases using these assumptions and parameters [20].

To begin, we assess mining task results based on their efficiency rather than their performance, so it is important to think about power usage [9]. Increasing memory clock speed improves performance, but increases memory error rate and energy consumption [8]. The most practical way to use this method is to increase the memory clock frequency. The clock frequency that offers the best results uses a disproportionate amount of energy and may be inefficient.

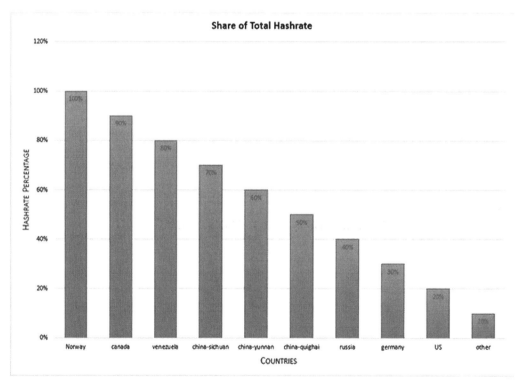

FIGURE 10.4
Hash rate share among the countries.

Mistakes may occur in mining task computations if overclocking is used in the mining process [12]. Due to the properties of the Ethash algorithm, the flow of mining errors for a particular value of the memory clock is fairly continuous. This can be empty or may contain a considerable fraction of the total number of Ethash function-generated values [5].

This will allow us to establish the mining clocks to their top limit while the memory clock rate is increased.

We can build a clock with the memory boost method associated with hardware overclocking for Ethash mining by combining the following facts:

1) For greatest efficiency, [13] a series of mining efficiency tests can be performed at a range of memory clock speeds, with the clock frequency increasing steadily in small increments.

2) Power consumption should be considered and efficiency calculated during the measurements.

3) Mining errors are taken into account when calculating efficiency. The experiment can be considered complete if the number of correctly calculated Ethash operations for a given clock frequency is smaller than the number of values of the properly calculated Ethash operations for a lesser frequency [15].

The selection of hardware is completely dependent on the miner and what cryptocurrency the miner wants to mine. The best hardware to mine bitcoin is Antminer S19 (see Table 1.1). Its power consumption is 3250W. If the miner overclocks the hardware, performance will increase, as will power consumption. The hardware will also produce much more heat than the standard temperature.

10.5 Advantages and Disadvantages of Mining Cryptocurrency

10.5.1 Advantages of Mining Cryptocurrency

Protection against inflation. Inflation has resulted in numerous coins reducing their value over time. Most cryptocurrencies are free at the time of their introduction with a fixed amount [16]. The ASCII text file indicates the quantity of every currency as there are 21 million bitcoins released around the world [18].

Administered and self-regulated. The administration and renewal of any currency are the greatest problems for its growth. Developers/miners store cryptocurrency group actions on their hardware and receive transaction fees as a reward [10]. Miners keep the transaction data correct, preserving the integrity of the cryptocurrency as well as the decentralized data.

Private and secure. Data protection has always been a serious challenge for cryptocurrencies. The blockchain ledger is predicated on extraordinary mathematical puzzles which require decoding [17]. This makes a cryptocurrency safer than normal digital exchanges. Pseudonyms are used to protect identities.

Currency exchanges are frequently administered virtually. Cryptocurrencies are usually sold in various currencies, just like the US dollar, the EU euro, the British pound, the Indian rupee, or the Japanese yen. With the help of various cryptocurrency wallets and transactions, one currency may be reborn in another by buying and selling in cryptocurrency through exceptional portfolios and with token transaction prices [14].

Decentralized. A *significant* aspect of cryptocurrencies is that they can be decentralized. An initial cryptocurrency operation is managed by developers, users of the cryptocurrency, and individuals who own a large amount of the currency, to increase it before it is discharged into the market [26]. Decentralization makes it possible to maintain a free monopoly of the currency so that no employer can claim the electricity for themselves, and on top of that, the really good value of the currency, which in turn can keep it strong and stable, is no longer like coin trust companies administered by the government [19].

Cash Value Transaction Mode. One of the keys that cryptocurrencies use is sending coins across borders. With the assistance of cryptocurrency, the exchange prices paid by an individual are reduced to a negligible amount [19]. By removing the requirement for third parties like VISA or PayPal to verify a deal, additional transaction costs are also eliminated.

Rapid money transfer. Crypto coins are continuously held thanks to the high-quality exchange solution [20]. Transactions, whether global or within the crypto home, are secured, which could be because of the very fact that the verification takes little or no time to action [5].

10.5.2 Disadvantages of Cryptocurrency Mining

Using Cryptocurrencies for Fraudulent Transactions. Since *the* privacy and security levels of cryptocurrency transactions are high, it is difficult for the government to find someone through their wallet or to track their information [27]. Bitcoin has been used as a way to exchange coins during illicit transactions on the dark web, for example, with narcotics [16].

Loss of statistics leads to financial loss. Developers have to make *near*-untraceable ASCII text files, strong hacking defenses, and unbreakable authentication protocols which can make cryptocurrencies more secure than physical cash or financial institutions [21]. But if someone loses their private key for their wallet, they cannot retrieve it. The wallet remains protected along with the number of coins it contains, which can lead to individual losses.

Decentralized but run by a couple of organizations. *Cryptocurrencies* are highly valued for their decentralization function. But the flow and number of cryptocurrencies within the trade are managed by their inventors and some organizations [22]. In this case, the holders control the currency themselves. Even highly indexed cash is subject to manipulation, such as bitcoin, the value of which doubled in 2017 [16].

Unable to take advantage of alternative fiat currencies. Some *cryptocurrencies* are better traded during a single currency or in some fiat currencies, forcing the user to convert those currencies into one of many key currencies such as bitcoin or Preliminary Ethereum, as their preferred currency through various exchanges [28]. Additional transaction prices overlap in the form of useless book values [8].

Result of cryptocurrency mining in the global environment. The development of crypto-mining has been extremely energy intensive. It requires superior, high-performance computer systems [23]. This is often impossible to achieve on standard computer

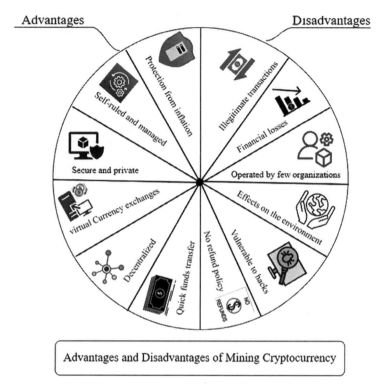

FIGURE 10.5
Advantages and disadvantages of mining cryptocurrency.

systems. Large bitcoin miners are in international locations like China, where coal is used to generate electricity, greatly increasing China's carbon footprint.

Vulnerable to hackers. Although cryptocurrencies are mostly stable, exchanges do not appear to be that stable. Most exchanges store customer wallet information [24]. This information may be stolen by hackers, who use it to access the primary business account. Exchanges, including Bitfinex, have been hacked within the past few years and large amounts of both bitcoin and dollars have been stolen [2]. Most of the exchanges are remarkably stable today, but there is always a chance of further deceptions.

Lack of cancellation or refund policies. If there is a *dispute* between parties, or someone sends money in error to the wrong wallet, the sender will not be able to get their coin back. Since there are no refunds, transactions can be created whose outcome or offer are never received (Figure 10.5).

10.6 Conclusion

This section demonstrates the need for a wider discussion on the environmental impacts of cryptocurrency mining and the amount of resource and electricity consumption that lies behind bitcoin.

While renewable energy is in intermittent supply, bitcoin makes relentless demands for electricity. Bitcoin ASIC mining hardware, once turned on, cannot be regenerated up to collapse or cannot mine bitcoin at a price. Bitcoin miners help to grow the demand for the baseline on a network. They use energy, not just as soon as there is more renewable energy available, but throughout the entire process of mining cryptocurrency.

10.7 Future Scope

Alongside the huge growth in cryptocurrencies in the past few years, many retailers and businesses have started accepting cryptocurrency for their products, for example Tesla Motors is accepting cryptocurrencies for purchase of electric vehicles. In the future, traditional currencies may be replaced by crypto currencies but this comes at a higher risk, as cryptocurrency mining consumes a lot of resources, whether electricity or hardware. There are moves to ban cryptocurrency mining in China which is one of the biggest cryptocurrency mining countries, because in China most of the electricity is produced from fossil fuels, with a detrimental effect on the environment.

References

1. Vujičić, D., Jagodić, D., & Ranđić, S. (2018, March). Blockchain technology, bitcoin, and Ethereum: A brief overview. In *2018 17th International Symposium Infoteh-Jahorina (Infotech)* (pp. 1–6). IEEE.
2. Williamson, S. (2018). Is bitcoin a waste of resources? Federal Reserve Bank of St. *Louis Review*, 100(2), 107–115. doi.10.20955/R.2018.107-15
3. Sukharev, P. V. (2020, January). Hardware overclocking to improve the efficiency of ethereum cryptocurrency mining. In *2020 IEEE Conference of Russian Young Researchers in Electrical and Electronic Engineering (EIConRus)* (pp. 1873–1877). IEEE.
4. Li, J., Li, N., Peng, J., Cui, H., & Wu, Z. (2019). Energy consumption of cryptocurrency mining: A study of electricity consumption in mining cryptocurrencies. *Energy*, 168, 160–168.
5. Sikorski, J. J., Haughton, J., & Kraft, M. (2017). Blockchain technology in the chemical industry: Machine-to-machine electricity market. *Applied Energy*, 195, 234–246. doi.10.1016/j.apenergy.2017.03.039
6. Naveen, N., & Thippeswamy, K. (2020). A Framework for Secure eHealth Data Privacy Preserving on Block chain with SHA-256 in Cloud Environment. *Turkish Journal of Computer and Mathematics Education (TURCOMAT)*, 11(3), 1118–1128.
7. Sovbetov, Y. (2018). Factors influencing cryptocurrency prices: Evidence from bitcoin, ethereum, dash, litcoin, and monero. *Journal of Economics and Financial Analysis*, 2(2), 1–27.
8. Sedlmeir, J., Buhl, H. U., Fridgen, G., & Keller, R. (2020). The energy consumption of blockchain technology: Beyond myth. *Business & Information Systems Engineering*, 62(6), 599–608.
9. Hindle, A. (2015). Green mining: A methodology of relating software change and configuration to power consumption. *Empirical Software Engineering*, 20(2), 374–409.
10. Ghimire, S., & Selvaraj, H. (2018, December). A survey on bitcoin cryptocurrency and its mining. In *2018 26th International Conference on Systems Engineering (ICSEng)* (pp. 1–6). IEEE.
11. Hayes, A. S. (2017). Cryptocurrency value formation: An empirical study leading to a cost of production model for valuing bitcoin. *Telematics and Informatics*, 34(7), 1308–1321. doi.10.1016/j.tele.2016.05.005

12. Mukhopadhyay, U., Skjellum, A., Hambolu, O., Oakley, J., Yu, L., & Brooks, R. (2016, December). A brief survey of cryptocurrency systems. In *2016 14th Annual Conference on Privacy, Security, and Trust (PST)* (pp. 745–752). IEEE.

13. Kugler, L. (2018). Why cryptocurrencies use so much energy: And what to do about it. *Communications of the ACM*, 61(7), 15–17.

14. Fadeyi, O., Krejcar, O., Maresova, P., Kuca, K., Brida, P., & Selamat, A. (2020). Opinions on sustainability of smart cities in the context of energy challenges posed by cryptocurrency mining. *Sustainability*, 12(1), 169.

15. Li, X., & Wang, C. A. (2017). The technology and economic determinants of cryptocurrency exchange rates: The case of Bitcoin. *Decision Support Systems*, 95, 49–60.

16. Ghosh, A., Gupta, S., Dua, A., & Kumar, N. (2020). Security of Cryptocurrencies in blockchain technology: State-of-art, challenges and future prospects. *Journal of Network and Computer Applications*, 163, 102635.

17. Cocco, L., & Marchesi, M. (2016). Modeling and Simulation of the Economics of Mining in the Bitcoin Market. *PloS One*, 11(10), e0164603.

18. Truby, J. (2018). Decarbonizing bitcoin: Law and policy choices for reducing the energy consumption of blockchain technologies and digital currencies. *Energy Research & Social Science*, 44, 399–410.

19. Hayes, A. (2015). A cost of production model for bitcoin. *SSRN Electronic Journal*, 1–4. doi.10.2139/ssrn.2580904

20. Calvão, F. (2019). Crypto-miners: Digital labor and the power of blockchain technology. *Economic Anthropology*, 6(1), 123–134.

21. Dorofeyev, M., Kosov, M., Ponkratov, V., Masterov, A., Karaev, A., & Vasyunina, M. (2018). Trends and prospects for the development of blockchain and cryptocurrencies in the digital economy. *European Research Studies Journal*, 21(3), 429–445.

22. O'Dwyert, K. J., & Malone, D. (2014). Bitcoin mining and its energy footprint. *IET Conference Publications*, 2014(CP639), 280–285. doi.10.1049/cp.2014.0699

23. de Vries, A. (2020). Bitcoin's energy consumption is underestimated: A market dynamics approach. *Energy Research & Social Science*, 70, 101721.

24. Hacioglu, U., Chlyeh, D., Yilmaz, M. K., Tatoglu, E., & Delen, D. (2021). Crafting performance-based cryptocurrency mining strategies using a hybrid analytics approach. *Decision Support Systems*, 142, 113473.

25. DeVries, P. D. (2016). An analysis of cryptocurrency, bitcoin, and the future. *International Journal of Buusiness Management and Commerce*, 1(2), 1–9.

26. Fadeyi, O., Krejcar, O., Maresova, P., Kuca, K., Brida, P., & Selamat, A. (2019). Opinions on sustainability of smart cities in the context of energy challenges posed by cryptocurrency mining. *Sustainability*, 12(1), 1–1.

27. Gallersdörfer, U., Klaaßen, L., & Stoll, C. (2020). Energy consumption of cryptocurrencies beyond bitcoin. *Joule*, 4(9), 1843–1846.

28. Azmoodeh, A., Dehghantanha, A., Conti, M., & Choo, K. K. R. (2018). Detecting crypto-ransomware in IoT networks based on energy consumption footprint. *Journal of Ambient Intelligence and Humanized Computing*, 9(4), 1141–1152.

29. Tahir, R., Huzaifa, M., Das, A., Ahmad, M., Gunter, C., Zaffar, F., Caesar, M., & Borisov, N. (2017). Mining on someone else's dime: Mitigating covert mining operations in clouds and enterprises. *Lecture Notes in Computer Science (Including Subseries Lecture Notes in Artificial Intelligence and Lecture Notes in Bioinformatics)*, 10453 LNCS, 287–310. doi.10.1007/978-3-319-66332-6_13

30. Phillip, A., Chan, J. S., & Peiris, S. (2018). A new look at cryptocurrencies. *Economics Letters*, 163, 6–9.

11

Cryptographic Hash Functions and Attack Complexity Analysis

Amit Bairwa, Vineeta Soni, Prashant Hemrajani, Satpal Singh Kushwaha, and Manoj Kumar Bohra

Manipal University, Jaipur, India

CONTENTS

11.1 Introduction

Computer security is all about the study and prevention of cyber-attacks. It is first necessary to investigate the principal motives behind such attacks. Hardly a month passes without some breaking cybersecurity news. Today, awareness of security policies and practices is a must for everyone. Consider a user account on a website. The most important aspect of it is how user passwords are protected. Any organization is at risk of database breaches through login credentials, so preventive measures are needed to ensure safety and protect user passwords [1]. This is where the concept of hashing comes into play. Instead of storing plain-text passwords directly in the database, we can store the password's hash. As attacks are continuously evolving, so too should advances in defence [2]. Defence should be active all the time, even when there is no attack. Types of attack include dictionary attacks, brute-force attacks, using look-up tables, reverse look-up tables, and rainbow tables.

Cryptography is based on three pillars: authenticity, integrity, and confidentiality [3]. Cracking a password is rather different from guessing it. Modern passwords are not stored in plain text in the database of user credentials [4]. Rather, they are stored in the form of a hash, which is produced by applying a hash function to the plain-text password [5]. It is not computationally possible (i.e., using mathematics) to recover the plain-text password from a hash. However, there are other ways in which it can be done. The simplest way is to guess the password, input it to a hash function, and compare the hash with the target hash [6]. This may seem easy to implement, but there are many hash functions out there. How can we tell what type of hash function is used in the particular hash to be cracked? If the hash is of the MD5 type, it must compute the MD5 hashes for all the password guesses till a match with the target hash is achieved, but if it is of the SHA512 type, the method must compute the SHA512 hashes of the password guesses. It would be easier to filter out hashes based on output length and compute hashes using only algorithms with the same output length [7]. If we have a 128-bit hash comprising 32 hexadecimal characters (each 4-bit), we know it could be an MD5, MD2, RIPEMD, or any other 128-bit hash, and we also know that this hash is not an SHA1, SHA224, SHA384, Tiger, Whirlpool, or other non-128-bit hashes [8]. Time and computational processing could be reduced if some hash computations were unnecessary. This can be done during a dictionary attack.

11.2 Brief Literature Review

A hash function can be visualized as a deterministic function that maps an input element from an infinite set to an output in a much smaller group [9]. For example, a cryptographic hash function h(x) maps an arbitrary length string to a fixed-length string depending upon the type of hash function. The one-way property of hash functions states that given a hash 'h' it is computationally infeasible to calculate 'x' such that h(x) = y. This property is one of the reasons why passwords are stored in the hash in the database, and if hashing is performed correctly and intelligently, they cannot be cracked. However, further measures are taken into consideration for security purposes (Rathod, Sonkar, and Chandavarkar 2020). The weak collision resistance property states that for a given input x1, it is computationally infeasible to find another input x2 such that h(x1) = h(x2). This property helps increase the complexity of any attack. If, for example, a brute-force attack occurs and the hash length is 'n' bits for the given input x1 [10], the total possible combinations would be 2n. If any string maps are allocated to any of the (2^n) combinations, it must generate a random string. Suppose x2, generate hash for it, and check if it is equal to h(x1) if not continue the loop, in the worst case this loop will run (O(2^n)) times. If the hash function has' possible outputs, the attacker needs 'N' guesses to find x2, such that h (x2) = h(x1) [7]. For the attacker to perform a brute-force attack to obtain input, he would have to loop his script as follows:

Suppose the given hash is y,
```
\emph{ do
generate a random string, x compute h(x)
while(h(x)! =y) return (x)
}
```

The substantial collision resistance property states that it is computationally infeasible to find two input values: h (x1) = h(x2). In weak collision resistance, the attacker would require (2^n) operations to succeed [11]. In strong collision resistance the attacker needs to perform $(2^{n/2})$ or sqrt$((2^n))$ operations to be successful, which is less than for weak collision resistance.

Attacking weak collision resistance is analogous to: "What is the minimum number of persons required in a class so that the probability of at least one person sharing a birthday with another person is greater than 0.5?" This is the same as asking what is the minimum number of inputs required for two inputs to have the same hash value so that the probability of 2 or more inputs having the same hash value is more significant than 0.5. Computing the minimum inputs required so that the (strong collision probability)\(>\)0.5 Suppose in a class of 'p' people we must calculate the probability such that any two people do not share their birthdays. The possibilities are:

A: two people among 'p' share their birthdays. B: no two people among 'p' share their birthdays.

$$P(A) = 1 - P(B)$$

Total number of ways to arrange 'p' people so that no one shares their birthday are:

$$\frac{365!}{p} * p! = \frac{365!}{p!(365-p)!} * p! \tag{11.1}$$

$\dfrac{365!}{p!(365-p)!} * p!$ is for choosing 'p' distinct days for 'p' people to avoid birthdays being on the same day.

and, p! is the total arrangements possible for p positions. The sample space would be $((365)^p)$. On simplification we get:

$$P(B) = \frac{365!}{p! * (365-p)! * (365) * P} \tag{11.2}$$

Now, on comparing these results with the strong collision terms, the probability of strong collision resistance (hash collision) is:

$$1 - \frac{n!}{(n-k)! * (n)^k} \tag{11.3}$$

If the hash length is of 'x' bits, then n = (2^x), that is, the total number of hashes possible, and 'k' is the total hashes the attacker can compute. This equation can further be reduced, as shown in Equation (11.4).

If n = (2^{64}) an k = (2^{32}), then p(hash collision) = 0.393 similarly after approximations conclude that if the total number of hashes possible are 'N' then sqrt(N) guesses are required to find x1 and x2 such that h(x1) = h(x2). Based on these properties, different hash functions can be compared to determine which are collision resistant and which cand be broken [12]. Some of the hash functions have proved that finding a collision is as difficult as some hard mathematical problems. They are called provably secure hash functions.

- A dictionary attack uses a word file that usually contains strings likely to be used as a password. Every word is hashed and is compared to the hash. The attacker guesses a word from the wordlist, hashes it, and compares [13].

- Brute-force attacks try every combination of characters up to a given length and compare its hash with the one which is to be cracked.
- Limitation: These kinds of attacks are very expensive computationally and take a lot of time if the password is long enough [14].
- Look-up tables are quite effective in cracking hashes. These methods pre-compute the hash value and store it corresponding to its plain text. Along with the strings, it also has the hash value for each entry. So, the hash to be cracked is searched in the table, and if found, its plain text is returned. The time complexity of this attack depends on how the implementation of the look-up table is done.
- The table contains md5 hashes, and the hash that is to be cracked, SHA1, would not work [15].
- Reverse look-up table attacks are used to exploit the vulnerability that two users can have the same passwords. Suppose the attacker has a compromised database, then the attacker maps each password hash from the database to the list of users having it. The attacker now hashes each password guess and searches for it in the look-up table. If found, then it is the list of users using that hash, i.e., users with the same passwords. This attack is effective because many users commonly have the same passwords. It can be used to get the users to have the same password [16].
- Rainbow-table attacks use rainbow tables; the concept is the same as look-up tables. The only difference here is that it stores a very small fraction of hash/password pairs. The idea is to organize the hashes and passwords so that data not kept can be recreated quickly. Increasing the cracking time and reducing the amount of memory represents a time/memory trade-off.

Nowadays, most hash-cracking platforms use massive look-up tables and take a fraction of a second to recover a password. Some typical platforms are Hashcat, John the Ripper, Crackstation.net, and Cain and Abel (a Windows password-cracking tool). The most popular and effective is Hashcat. It offers more than 300 supported hash functions and many types of attacks such as straight, combination, brute-force, hybrid, and dictionary attacks. Brute-force attacks are used when all else fails as they are the most time-consuming and computationally expensive with exponential time complexity [17]. Rule-based attacks are complicated, and they are like a programming language for password candidate generation [18]. This type of attack can be used if easier ones fail and the person trying to recover the password has some idea of the target. Hashcat can also utilize the power of GPU to speed up the attacks, making them one of the world's fastest password crackers.

11.3 Comparison of MD5 and SHA1 Based on Collision-Resistant Property

Collision resistance is the property of a hash function that makes it computationally infeasible to find two colliding inputs. Different hash functions can be compared based on their collision probability. This collision probability can be thought of as analogous to the birthday paradox. With the increase in the number of possible output combinations, the probability of hash collision decreases on the same number of hash computations or guesses

TABLE 11.1

MD5 Hash Function

Number of Hashes Attacker Computes	Collision Probability (Approx.)
2	0
2^{30}	0
2^{62}	0.038
2^{63}	0.117
2^{64}	0.393
2^{66}	0.999

by the attacker or any other person. For example, in an MD5, the hash output length is 128 bits, but in an SHA1 hash, the output length is 160 bits, which means the total possible combinations are more than MD5. Suppose an attacker can compute hashes. A hash collision would probably occur, which would be approximately 0.393 in the case of an MD5 hash function. In contrast, in the SHA1 hash function, the probability of hash collision would reduce to around 0 (Tables 11.1 and 11.2).

It can be clearly observed that if an attacker computes 2^{64} MD5 hashes, there is a 40% chance of hash collision, whereas the attacker needs to compute 2^{80} SHA1 hashes to generate a 40% chance of collision. Hence, $\sqrt{N}$ computations would be required to generate a decent chance of collision. where N is the total possible hash combinations. So the longer the output, the more computations will be required to generate collisions. If any collisions are found before computing $\sqrt{N}$ hashes, then those hash functions are considered broken.

$$1 - e^{\frac{-k(k-1)}{2N}} \tag{11.4}$$

11.4 Analysis of Dictionary Attack

Wordlists of arbitrary size can be used while performing this attack. The processing time taken to match the word with the target hash correctly depends on the length of the wordlist. When a given hash is to be cracked, (a) a random word is guessed from the wordlist, (b)

TABLE 11.2

SHA1 Hash Function

Number of Hashes Attacker Computes	Collision Probability (Approx.)
2	0
2^{40}	0
2^{78}	0.038
2^{79}	0.117
2^{80}	0.393
2^{82}	0.999

TABLE 11.3

Cracking a MD5 Hash Using a 10,000-Word Wordlist

Attempts	Time (Seconds)
7604	0.0161
5392	0.0102
421	0.0021
19416	0.0367
12015	0.0319

TABLE 11.4

Cracking a MD5 Hash Using a 14,400,000-word Wordlist

Attempts	Time (Seconds)
1137650	2.6903
511015	1.0714
47234	0.1076
260407	0.5351
1216799	2.619

hashing is applied to it, and whether it is the same as the target hash is checked, then (c) if the target string and guessed word are not identical, the whole process must be repeated.

11.5 Observations

The method uses password lists of different sizes to try to crack the hash. 'Attempts' mean the number of words that are guessed until the word is found whose hash matches the target hash. The number of attempts is up-to-the random word generator and is unpredictable. Tables 11.3 and 11.4 list observations with an Intel(R) Core (TM) i7-8550U CPU @ 1.80 GHz processor and 8 GB RAM.

11.6 Password Storage Concepts: Salting

Salt is a random string that is concatenated with passwords before being operated on by a one-way function. Randomization of hashes is possible. Suppose two users have the same passwords and salting is not done. The same hash value would be stored in the database. If the database is leaked, an attacker could efficiently run a reverse lookup-table attack to find users with the same hash.

In contrast, if salting is done, a different hash value would be generated for the same password. So randomizing the hash lookup tables renders reverse lookup tables and rainbow tables ineffective. The salt value is usually stored in the database with the hash itself,

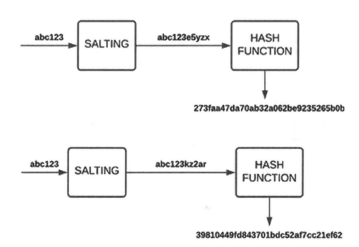

FIGURE 11.1
Same password having different hash values.

as it is required when a user tries to log in. Salt should be generated using a cryptographically secure pseudo-random number generator (CSPRNG). Salt should not be reused, and a short salt length should be avoided. Figure 11.1 shows two users having the same password, "abc123," but different hash values would be stored in the database with the salt due to salting. Random strings are concatenated after the original password in both cases. A weak password is used in this example for simplicity, and the hash function used is MD5.

11.7 Statement of the Problem

The present research focuses on cryptographic hash functions and the complexity of various attacks used to crack hashes. Building a simple and easy-to-use command-line tool that would hash any arbitrary length input, also applies to salt. The plain text can be retrieved if required if the input hash is provided using pre-computed hash tables, referred to as lookup tables. This method could reduce the time taken and the CPU usage. Here we compare and evaluate various attacks based on their complexity.

11.8 Objectives of the Study

This study aims to improve understanding of cryptographic hash functions, properties of these functions, types of hash function, and password hashing. How to hash properly, how hashes are cracked, various possible attacks on hashes, adding salt to improve security, and ineffective hashing methods are discussed. A command-line tool based on hashing is developed to compare different hash functions based on their security, complexity, and collision-resistant properties (Figure 11.2).

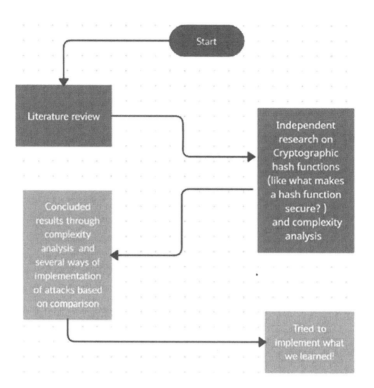

FIGURE 11.2
Research methodology.

11.9 Results and Discussion

Attacks have different levels of complexity. The most basic attacks are the brute-force and dictionary attacks. The brute-force attack, which is easy to implement, tries every possible combination of characters up to a given length. These attacks are costly, and there is no way to prevent them. If a password hashing system is robust, then only brute-force and dictionary attacks are possible, which can be specified as a complement to the password hashing system (Figures 11.3 and 11.4).

11.10 Analyzing the Complexity of Brute-Force Attacks

Suppose there are 'n' possible characters using which password combinations can be formed. The possible number of passwords with different password lengths are listed in Table 11.5.

So, for 'k' password length, a total of n k passwords is possible. Therefore, the brute-force attack has a time complexity of ($O(n^k)$). Dictionary attacks work by guessing a word from the wordlist, hashing it, and then comparing it with the hash which is to be cracked.

FIGURE 11.3
Brute-force attack.

```
Searching: 5f4dcc3b5aa765d61d8327deb882cf99: FOUND: password5
Searching: 6cbe615c106f422d23669b610b564800: not in database
Searching: 630bf032efe4507f2c57b280995925a9: FOUND: letMEin12
Searching: 386f43fab5d096a7a66d67c8f213e5ec: FOUND: mcd0nalds
Searching: d5ec75d5fe70d428685510fae36492d9: FOUND: p@ssw0rd!
```

FIGURE 11.4
Lookup-table attack.

TABLE 11.5

Password Combinations With Brute-Force Attacks

Password Length	Number of Passwords Possible
1	n
2	n^2
3	n^3
4	n^4
.	.
k	n^k

Suppose the wordlist contains 'n' words, in the worst case, if the hash matched is (n-1) the word, the time complexity of this attack would be O(n) or O (size of the wordlist). Lookup tables are an extremely effective method for cracking many hashes of the same type very quickly. The basic idea behind these tables is to pre-compute the hashes of a dictionary containing passwords and store them. This is done in the form of a lookup table data structure. The complexity of the lookup tables depends on how they have been implemented. One way is for hashes for all the words in the dictionary file (ex: rockyou.txt) for all the hash functions to be stored as a key-value pair. When the process has the target hash, it will search for it in the hash map and, if found, the word will be returned.

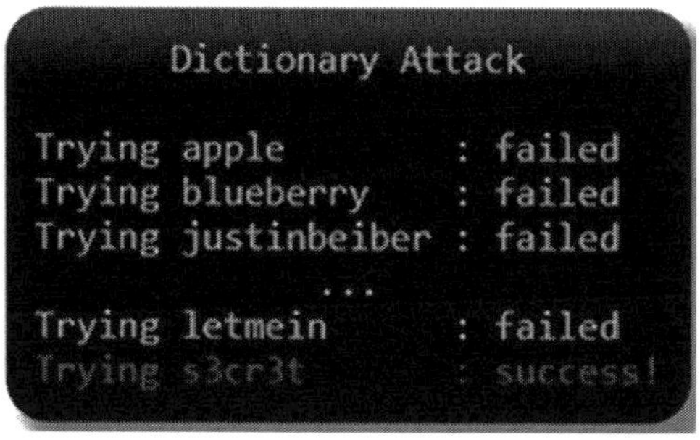

FIGURE 11.5
Dictionary attack.

However, this would be a very tedious task, as the hashes of every function must be stored in a single file. There are usually 14.3M words, so if the method has 'n' hash functions, then the total number of words will be (14.3 * n) M words. It's going to be very large. We know that searching in a hash-map or a dictionary takes O (1) time, but it would be better to have separate lookup tables for hashes as handling this large file can cause some issues. So, it would be better to have n lookup tables for each hash function. So now the issue is: how can it search in separate lookup tables?

When the target hash is known, instead of searching for it in all lookup tables, the tables could be further filtered. The length of the hash which is to be cracked can be computed, so if it turns out to have 128 bits, then MD5, RIPEMD and other 128-bit hashes are possible. So now, it will only search for the hash in the lookup tables for these algorithms, instead of searching in all of them. Let's take an example of a hash (Figure 11.5):

9b71d224bd62f3785d96d46ad3ea3d73319bfbc2890caadae2dff72519673ca7

2323c3d99ba5c11d7c7acc6e14b8c5da0c4663475c2e5c3adef46f73bcdec043

Here, this hash is 512 bits, so that it could be SHA512, SHA3512, Blake2b, or other 512-bit hashes. After looking for it in the selected lookup tables, it will receive the corresponding plaintext as "hello," which means it has been cracked. While it takes time to compute the lookup tables, the lookup operation itself takes a fraction of a second.

11.11 Conclusion and Future Work

This work has extended previous work by applying it in a study of the three functions: generating the hash, salting, and cracking the hash. A command-line script generates the hash based on user input, salting can be used if needed, and the hash is cracked using a look-up table. The first two functions have been successfully implemented, but there remains some work to get the third function completed. When complete, the implementation is planned to be open source, to make it available publicly and invite further development or modifications if needed.

References

1. M. S. Al-Ali, H. A. Al-Mohammed, and M. Alkaeed, "Reputation based traffic event validation and vehicle authentication using blockchain technology," *2020 IEEE International Conf. Inform. IoT Enabling Tech. (ICIoT)*. IEEE, pp. 451–456, 2020.

2. W. I. Khedr, H. M. Khater, and E. R. Mohamed, "Cryptographic accumulator-based scheme for critical data integrity verification in cloud storage," *IEEE Access*, vol. 7, pp. 65635–65651, 2019, doi: 10.1109/ACCESS.2019.2917628

3. P. R. Raju, and K. C. Shet, "A cryptographic hashing solution for mitigating persistent packet reordering attack in wireless ad hoc networks," *Proc. Turing 100 - Int. Conf. Comput. Sci. ICCS 2012*, pp. 379–383, 2012, doi: 10.1109/ICCS.2012.83

4. N. Abdoun, S. El Assad, M. A. Taha, R. Assaf, O. Deforges, and M. Khalil, "Secure hash algorithm based on efficient chaotic neural network," *IEEE Int. Conf. Commun.*, vol. 2016-August, pp. 405–410, 2016, doi: 10.1109/ICComm.2016.7528304

5. E. Böhl, M. Lewis, and K. Damm, "A collision resistant deterministic random bit generator with fault attack detection possibilities," *Proc. - 2014 19th IEEE Eur. Test Symp. ETS 2014*, pp. 14–15, 2014, doi: 10.1109/ETS.2014.6847829

6. M. Mozaffari-Kermani, and R. Azarderakhsh, "Reliable hash trees for post-quantum stateless cryptographic hash-based signatures," *Proc. 2015 IEEE Int. Symp. Defect Fault Toler. VLSI Nanotechnol. Syst. DFTS 2015*, pp. 103–108, 2015, doi: 10.1109/DFT.2015.7315144

7. M. H. N. Ilahi, M. Syahral, and B. H. Susanti, "Collision attack on 4 secure PGV hash function schemes based on 4-round PRESENT-80 with iterative differential approach," *2019 16th Int. Conf. Qual. Res. QIR 2019 - Int. Symp. Electr. Comput. Eng.*, 2019, doi: 10.1109/QIR.2019.8898295

8. D. Z. Sun, and J. D. Zhong, "A hash-based RFID security protocol for strong privacy protection," *IEEE Trans. Consum. Electron.*, vol. 58, no. 4, pp. 1246–1252, 2012, doi: 10.1109/TCE.2012.6414992

9. W. Yu, and Y. Jiang, "Mobile RFID mutual authentication protocol based on hash function," *Proc. - 2017 Int. Conf. Cyber-Enabled Distrib. Comput. Knowl. Discov. CyberC 2017*, vol. 2018-January, pp. 358–361, 2017, doi: 10.1109/CyberC.2017.45

10. A. Ali Alkandari, I. F. Al-Shaikhli, and M. A. Alahmad, "Cryptographic hash function: A high level view," *Proc. - 2013 Int. Conf. Informatics Creat. Multimedia, ICICM 2013*, pp. 128–134, 2013, doi: 10.1109/ICICM.2013.29

11. P. Assamongkol, S. Phimoltares, and S. Panthuwadeethorn, "Enhancing trustworthy of document using a combination of image hash and cryptographic hash," *Proc. 2017 14th Int. Jt. Conf. Comput. Sci. Softw. Eng. JCSSE 2017*, pp. 1–6, 2017, doi: 10.1109/JCSSE.2017.8025924

12. H. P. Patel, and M. B. Chaudhari, "A time space cryptography hashing solution for prevention jellyfish reordering attack in wireless adhoc networks," *2013 4th Int. Conf. Comput. Commun. Netw. Technol. ICCCNT 2013*, 2013, doi: 10.1109/ICCCNT.2013.6726689

13. K. Aggarwal, and H. K. Verma, "Hash-RC6 - variable length hash algorithm using RC6," *Conf. Proceeding - 2015 Int. Conf. Adv. Comput. Eng. Appl. ICACEA 2015*, pp. 450–456, 2015, doi: 10.1109/ICACEA.2015.7164747

14. N. Sklavos, "Towards to SHA-3 hashing standard for secure communications: On the hardware evaluation development," *IEEE Lat. Am. Trans.*, vol. 10, no. 1, pp. 1433–1434, 2012, doi: 10.1109/TLA.2012.6142498

15. C. Bouillaguet, P. Derbez, O. Dunkelman, P. A. Fouque, N. Keller, and V. Rijmen, "Low-data complexity attacks on AES," *IEEE Trans. Inf. Theory*, vol. 58, no. 11, pp. 7002–7017, 2012, doi: 10.1109/TIT.2012.2207880

16. F. Legendre, G. Dequen, and M. Krajecki, "Encoding hash functions as a SAT problem," *Proc. - Int. Conf. Tools with Artif. Intell. ICTAI*, vol. 1, pp. 916–921, 2012, doi: 10.1109/ICTAI.2012.128

17. J. Li, G. Zeng, and Y. Yang, "Collision attack framework on RIPEMD-128," *Proc. - 2020 2nd Int. Conf. Artif. Intell. Adv. Manuf. AIAM 2020*, pp. 321–324, 2020, doi: 10.1109/AIAM50918.2020.00071

18. Ai Gu, Z. Yin, C. Cui, and Y. Li, "Integrated functional safety and security diagnosis mechanism of CPS based on blockchain," *IEEE Access*, vol. 8, pp. 15241–15255, 2020, doi: 10.1109/aCCESS.2020.2967453

19. P. Gauravaram, "Security analysis of salt||password hashes," *Proc. - 2012 Int. Conf. Adv. Comput. Sci. Appl. Technol. ACSAT 2012*, pp. 25–30, 2012, doi: 10.1109/ACSAT.2012.49

20. K. Quist-Aphetsi, and H. Blankson, "A hybrid data logging system using cryptographic hash blocks based on SHA-256 and MD5 for water treatment plant and distribution line," *Proc. - 2019 Int. Conf. Cyber Secur. Internet Things, ICSIoT 2019*, pp. 15–18, 2019, doi: 10.1109/ICSIoT47925.2019.00009

12

Mixed Deep Learning and Statistical Approach to Network Anomaly Detection

Vineeta Soni, Siddhant Saxena, Devershi Pallavi Bhatt, Narendra Singh Yadav, Amit Bairwa, and Satpal Singh Kushwaha
Manipal University, Jaipur, India

Sarada Prasad Gochhayat
Virginia Modeling, Analysis, and Simulation Center, Old Dominion University, Suffolk, Virginia, USA

CONTENTS

12.1 Introduction

Detection of anomalous packets in network traffic in a zero-day attack situation is complex because organizations such as banks and hospitals rely on traditional methods [1].

An alternative approach uses deep learning and supervised training [2] on a dataset containing both normal packet information and various types of anomalies in the form of feature-value mapping. However, attacks on different institutions vary, and training that uses data containing anomaly labels on specific types of security attack cannot perform accurately in the case of a zero-day attack.

This chapter describes an approach to classifying normal packets and anomalous packets that uses a deep learning model trained on a large feature-value mapping dataset of normal network traffic packets directly captured from an organization's network traffic. The model learns its weights and biases using backpropagation and updates the accuracy of its results using probabilities from a multilayer perceptron model and a statistical probability distribution model.

12.2 Network Anomaly Detection

The process of detecting data patterns that are different from the regular flow of data, when the normal flow of a network is attacked by an entity, is called network anomaly detection [3].

Traditional machine-learning approaches to anomaly detection use supervised learning methods based on available large-scale datasets such as KDD Cup. The main problem with the traditional supervised approach is the growth of newly emerging attack methods on which the models are not trained. Data preparation for training the detection model is also very difficult because normal network traffic can be captured easily, but the architecture of attacks varies for different types of institutions, and collecting data by simulating attacks to train the detection algorithm is a costly operation.

A supervised approach is therefore used, in which a deep learning model is trained on collected normal network data, so if any type of attack or abnormality occurs, the intrusion will automatically be labelled.

12.3 Traditional Approaches to Network Anomaly Detection

Table 12.1 describes machine-learning algorithms for detecting anomalies in network packets [4]. A KDD 199 Cup dataset is used, which contains millions of records labelled as normal or as a particular kind of intrusion. Each record contains 41 features. Intrusions present in the dataset are of four main kinds—dos, probe, r2l, u2r—with 24 sub-types in the training set and 14 additional sub-types in the test set only.

12.4 Deep Learning Model Flow

The first step in training a deep learning model to perform anomaly detection is to collect data. Data is collected in a raw format by tapping the network traffic between client,

TABLE 12.1

Traditional Machine-Learning Algorithms for Network Anomaly Detection

Algorithm Description	Methodology Comparison	Results
Decision Tree [5]: Supervised learning method used for classification. Uses an algorithmic approach that identifies ways to split data sets based on different conditions.	A decision tree selects the most important features for building a tree and then performs pruning operations to remove irrelevant branches from the tree to avoid overfitting the model. Ex. XGBoost	Leaf size = 50, Precision: (a) 0.764712 for normal/anomaly (b) 0.72.5842 for types of anomalies
Random Forest [6]: Classification algorithm consists of many decision trees. Uses techniques like feature randomness and bagging, and tries to create a forest of trees whose prediction is better than a single tree.	Ensemble classification approach in which a random subset of features is considered for splitting a node. Efficient with a low false alarm rate and high detection rate.	Number of trees = 500 Precision: (a) 0.769221 for normal. anomaly (b) 0.705164 for types of anomalies
Naive Bayes [7]: Classification algorithm based on Bayes' Theorem, which assumes every feature present in a dataset is not correlated with other features.	Makes hard assumptions regarding independence, but helps to find important features using correlation, information gain, and gain ratio.	Precision: (a) 0.772576 for normal/anomaly (b) 0.496871 for types of anomalies
K Nearest Neighbour (KNN) [8]: Supervised learning algorithm which stores all available data, points and classifies new data by a majority vote of its nearest data points.	Based on feature similarity to predict the class of a certain data point and a comparison between values of k to identify best-fit model using benchmark datasets. Ex: CSE-CIC-IDS2018	$K = 30$ Precision: (a) 0.749218 for normal/anomaly (b) 0.69.701643 for types of anomalies

server, and an open internet connection. This data—just network packet information—is in raw form. The data is processed and the features that are most useful for training models are extracted and converted to tensor objects. This step is called data preprocessing. The multilayer perceptron network takes input in the form of tensors and learns by mapping weights and biases. It uses a backpropagation algorithm to update them; outputs of the probabilities of the packet are either normal or anomalous. Figure 12.1 shows the step-by-step block process of collecting data and using it to train the multilayer perceptron model.

12.5 Preparation of Dataset by Tapping Network Traffic

Network traffic is defined as data packets moving across a network at a given point in time. Network packets provide the load in the network, and require analysis to detect anomalies, since any type of attack/malicious activity occurring during packet transactions also reaches the client device. In order to collect network intrusion data, the network and packets with their information are tapped using Tcpdump and Wireshark respectively.

Raw .pcap file data contains all forms of internet transaction packets, which can be segregated into small .pcap files related to different types of data, exchange packets using ports, as explained in Figure 12.2.

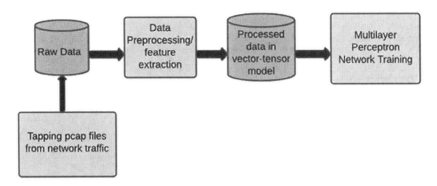

FIGURE 12.1
Deep learning workflow chart.

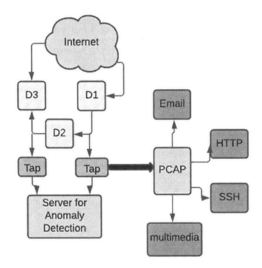

FIGURE 12.2
Data extraction in pcap format files.

Pcap is an application programming interface (API) for capturing packets in the network to which the computer is attached .Pcap files can be filled with captured data using the Tcpdump Packet Analyzer, a computer program that runs under a command-line interface. For example, to capture 100 packets from any interface in a Linux environment, use the command:

```
$ sudo tcpdump -i any -c 100
```

.pcap files can be analyzed using the Wireshark tool.

The raw .pcap format file contains all types of packet transfer in different types of transaction such as email, http, SSH, and multimedia [10].

These files are used to extract features and their values, and are converted into a comma-separated file which serves as input to calculate joint probabilities of features that can be used as test files, and to calculate predictions from a multilayer perceptron model. For an effective performance of the overall detection model, the large raw-data .pcap file is split

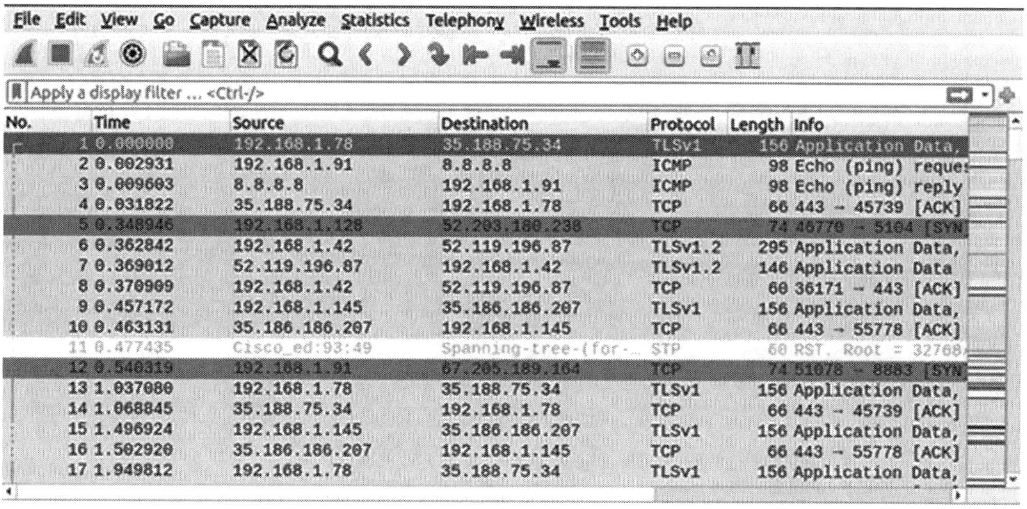

FIGURE 12.3
Data file analysis using Wireshark.

into small .pcap files with respect to labels and protocols, the features of each file are extracted, the model is trained on them, and the results are combined.

12.6 Analysis of Tapped Network Packets Using Wireshark

Packets captured using Tcpdump can be analyzed using the Wireshark graphical interface tool which visualizes all packets, their source and destination IP addresses, their protocol, length, and descriptive information (Figure 12.3), captures random packets and dumps them into a .pcap format file [11]. Figure 12.4 shows features and their mapping values for every packet that can be extracted from protocol information.

12.7 Feature Extraction from a .pcap File

Feature extraction reduces raw data, converting it into meaningful features so that the relation between feature and value mapping in the dataset can be used to train models fast and efficiently. Transaction types occurring in the network include emailing, HTTP sessions, SSH sessions, and multimedia. Various application-layer protocols for sending and receiving mails are available, and the combination of these protocols helps with an end-to-end email exchange between users. The three most commonly used application-layer protocols are POP3 (Post Office Protocol 3), IMAP (Internet Mail Access Protocol), and SMTP (Simple Mail Transfer Protocol).

Selection of features from the raw .pcap file is a very important aspect of training a deep learning model [13], because it reduces the complexity of a model, enables the algorithm to train faster, and makes it easier to interpret.

```
▾ Transmission Control Protocol, Src Port: 45739, Dst Port: 443, Seq: 1, Ack: 1, Len: 90
    Source Port: 45739
    Destination Port: 443
    [Stream index: 0]
    [TCP Segment Len: 90]
    Sequence number: 1     (relative sequence number)
    Sequence number (raw): 2221710259
    [Next sequence number: 91     (relative sequence number)]
    Acknowledgment number: 1     (relative ack number)
    Acknowledgment number (raw): 2459008662
    1000 .... = Header Length: 32 bytes (8)
  ▸ Flags: 0x018 (PSH, ACK)
    Window size value: 2529
    [Calculated window size: 2529]
    [Window size scaling factor: -1 (unknown)]
    Checksum: 0x833f [unverified]
    [Checksum Status: Unverified]
    Urgent pointer: 0
  ▸ Options: (12 bytes), No-Operation (NOP), No-Operation (NOP), Timestamps
  ▸ [SEQ/ACK analysis]
  ▸ [Timestamps]
    TCP payload (90 bytes)
```

FIGURE 12.4
The description of a packet using Wireshark.

Eighteen features are extracted from the .pcap files and each .pcap file containing packet transaction information on a particular type of protocol is used as input in the form of the vector-tensor model.

12.8 Feature Selection

Features are the categories of each data packet that distinguish it from other data packets. Prediction of the results depends heavily on the selection of features [14].

The important features are destination port, TCP flags, protocol, and IP header length. These features have very low correlation values and can be regarded as independent from each other; hence they also provide a better statistical analysis of the normal distribution of a particular network's packet flow.

12.9 Extracting Features from Raw .pcap File

To extract the feature-value mapping, the CyberSecTK tool [15] can be used. It uses Python libraries called OS and Glob to filter raw .pcap files to category-wise files based on the protocols and ports used by packets in network traffic. Table 12.2 shows how each filter is named for a general understanding of categories; here the filtering criteria column contains the '&& port' protocol, and the operation is a logical 'and' operation.

After segregating a large raw .pcap file into small files containing packet types that are transferred into respective categories, each packet's information is extracted from all these .pcap files and converted into CSV format files having features and their respective value mapping.

TABLE 12.2

Filtered Files Name and Filtering Criteria

Filtered File Name	Filtering Criteria
Email_file	smtp && (smtp.port == 25)
HTTP_file	http && (http.port == 80)
SSH_file	tcp && (tcp.port == 22)
Multimedia_file	sctp && udp && (udp.port == 68)

TABLE 12.3

Illustration of an Anomalous Packet in the Packet Transaction Flow

Packet Number	Size of Payload	Destination Port
P1	2	22
P2	10	400
P3	20	80

12.10 Statistical Analysis of Data Set

Data extracted from a raw .pcap file and turned into a vector-tensor model have a column-wise value to feature mapping, and when we plot the data with respect to feature it forms a Gaussian distribution curve. Table 12.3 is an example of normal data and anomalies that demonstrate variation in values between features of normal and anomalous packets.

Packets that have no anomalies, (i.e., normal packets lying in the middle part of the Gaussian distribution curve) have some standard deviation from the mean but packets that contain anomalous properties differ by a large number of values corresponding to features and tend to show a large deviation from the standard curve (see Figure 12.5). The probability of a packet given mean and standard deviation can be used to specify whether it lies in normal distribution or contains anomalous properties [16].

12.11 Statistical Anomaly Detection Using Joint Probability Approach

The probability of each packet in a given feature with respect to the given mean and variance can be calculated using Equations (12.1) and (12.2).

$$\text{mean}(v) = \frac{1}{n}\sum_{i=1}^{n}x_i \tag{12.1}$$

$$\text{standard deviation}(\sigma^2) = \sum_{i=1}^{n}(x_i - v)^2 \tag{12.2}$$

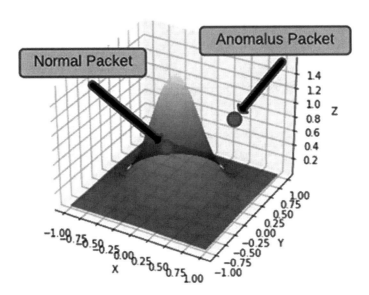

FIGURE 12.5
Gaussian distribution plot. Red points indicate normal and anomalous packets.

The probability distribution of a packet given the single feature vector can be calculated using mean and variance as shown in Equation (12.3).

$$P\left(x,v,\sigma^2\right)=\left(\frac{1}{\sqrt{2\Pi}\sigma}\right)e^{-\frac{\left(x_j-v_j\right)^2}{2\sigma_j^2}} \tag{12.3}$$

But to calculate the probability of a packet being normal or anomalous, the joint probability (Equation 12.4) is calculated by multiplying all probabilities assuming all features are independent of each other.

$$P\left(X\right)=P\left(x1\right)*P\left(x2\right)*\ldots\ldots*P\left(xn\right) \tag{12.4}$$

Hence, the overall probability score of a particle can be calculated using Equation (12.5).

$$P\left(x\right)=\prod_{j=1}^{n}\left(\frac{1}{\sqrt{2\Pi}\sigma_j}\right)e^{-\frac{\left(x_j-v_j\right)^2}{2\sigma_j^2}} \tag{12.5}$$

This probability of a packet is compared to ϵ(variable), which is calculated using the trial-and-error method. If $P(x) > \epsilon$, it is regarded as a normal packet, otherwise the packet is labelled anomalous.

However, this statistical approach [17] is not very efficient in distinguishing a packet as being anomalous because of the hard assumption made when analyzing each of the packet's features, independent of all other features, to calculate joint probabilities. The real-world dataset of any collection of packets contains some features that are highly correlated with each other, and some that are less so. Hence labelling of anomalies solely on the basis of the P(x) calculation is not sufficient for achieving anomaly detection with a good level of precision.

12.12 Multilayer Perceptron Model

The multilayer perceptron (MLP) model corresponds to a feed-forward neural net [18], which has an input layer, one or more hidden layers, and an output layer. It can be described as a whole network with a non-linear transformation of feature values learned through training with the help of hidden layers. MLPs are fully connected, with each node in one layer connected to every node in the next layer with a certain weight ($w_{j,j}$). Data present in feature-value mapping can be converted to tensors which are inputs to the neuron, are then multiplied by their corresponding synaptic weights and a bias (b) term is added to them, generating activation potential (z).

$$\text{activation potential}(z) = \sum_{j=1}^{n} x_j w_{j,j} + b \tag{12.6}$$

Then z is passed to a non-linear activation function as the output of the packet's probability being anomalous or normal lies in the range 0 to 1. The sigmoid activation function, used to limit output values of neurons, is defined in Equation (12.7):

$$sigmoid(x) = \frac{1}{1 + e^{-x}} \tag{12.7}$$

Each neuron in all the hidden layers consists of this linear activation potential and non-linearity (sigmoid function). Figure 12.6 represents the structure of a single neuron, which is connected by the input layer through weights and output via activation potential. In Figure 12.6, a_j is the perceptron, which is a combination of activation potential (z) and non-linear sigmoid function (g). The weights between each layer are grouped with a matrix W, so activation potential z can be written in a compact form as

$$z = W^T x.... \tag{12.8}$$

12.13 Architecture of Multilayer Perceptron

An MLP model [19] can take any number of inputs and can be used to predict any number of output classes as shown in Figure 12.7. Here layers ($l = 1$ and $l = 2$) are hidden layers, and layer ($l = 0$ and $l = 3$) are input and output layers respectively.

The strength lies in connecting these layers with respect to their weights and biases, therefore the whole network can be described with a non-linear transformation that uses a combination of Equations (12.6) and (12.7).

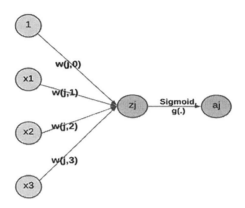

FIGURE 12.6
The structure of a single neuron having activation potential.

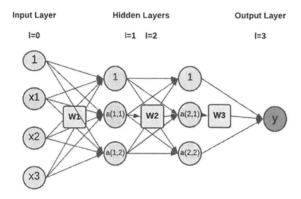

FIGURE 12.7
Multilayer perceptron model.

Recursively, each layer (*l*) can be written as

$$a^{(l)} = g\left(w^T a^{(l-1)}\right),\dots \tag{12.9}$$

with $a^{(0)} = x$ being the input layer to the network and $a^{(L)}$ being the output.

An MLP is trained by updating weights after each batch of data is processed, based on loss calculation, which is an error in the model output compared to the basic ground truth provided. Training is a supervised task in which the model learns a function through a backpropagation algorithm based on input–output mapping pairs. The loss function used to calculate the error or accuracy score can be the least mean square method, binary cross-entropy loss, etc.

12.14 Binary Classification of Labels Using PyTorch

PyTorch is an open-source framework to perform machine learning, computer vision, and natural language processing (NLP) based on the Torch library.

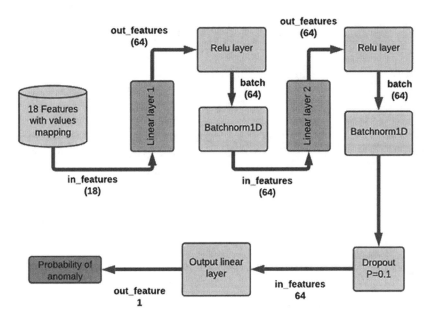

FIGURE 12.8
Architecture of multilayer perceptron model.

PyTorch torch models are imported for a multi-perceptron deep neural network [20], using torch.nn modules. Adam optimization is the most suitable for a binary classification of anomalies; for input–output operations several utilities under the utils class module from which a data loader can be imported are present.

Eighteen features prepared from the raw .pcap file derived from the dataset are input to the linear layer. Similarly ReLU [21], Dropout, and BatchNorm layers are defined (see Figure 12.8).

The sigmoid activation function is not used in the final layer during training because the BCEWithLogitsLoss function automatically applies the sigmoid activation.

12.15 Architecture of the Multilayer Perceptron Model for Anomaly Classification

To implement anomaly classification [22], the Classification class label is used which imports nn. module class from PyTorch. In the init function, the layers to be used can be specified.

12.16 Improving Model Accuracy Using Statistical Prediction of Anomalies

The joint probability approach provides a probability for each packet and compares it with a threshold calculated using the trial-and-error method. Anomalies can be detected with some precision, but the major problem is to calculate probabilities for all particles.

Although features are regarded as fully independent, this is not the case in practice, hence it is not the best way to calculate the probability.

To train a deep learning model [23] to predict with good precision requires a backpropagation algorithm to compute losses and update weights and biases.

When raw data is extracted using the .pcap files, features and value mapping can be performed, but there are no labels to validate the accuracy and to update weights.

Hence, statistical prediction is used as an evaluation parameter [24]. The probability of each packet can be stored in a list and the training loop can evaluate the model's prediction using the deep learning and after every prediction, weights and biases are updated using loss, which is based on comparison of probabilities, outputs of statistical distribution and the multilayer perceptron model.

In Algorithm 12.1, *model_accuracy, y_pred* and *y_test* are inputs for a batch (here, 64), where y_pred and y_test are lists containing predictions generated using the deep learning model and statistical probability distributions, respectively.

Algorithm 12.1: Multilayer Perceptron Accuracy calculation

```
Step 1: Start
Step 2: Read input lists y_pred, y_test.
Step 3: Roundoff the y_pred and y_test lists using torch.round()
Step 4: Declare variable results
        4.1: results ← (y_pred == y_test).sum().float()
Step 5: Declare variable accuracy
        5.1: accuracy ← results/y_test.shape[0]
        5.2: Roundoff ( accuracy * 100 ) using torch.round()
Step 6: Stop
```

Both probabilities are rounded off to convert them into either a 0 or a 1, then the model accuracy can be calculated.

12.17 Training the Deep Learning Model Using PyTorch

A mixed approach to training the deep learning model is used. The specific PyTorch terms used for training are as follows:

- Model.train() :-
  Train_loader provides an iterable over the given dataset, looping can be used to get data in batches from the train_loader.
- Optimizer.zero_grad() :-
  Restarts looping without losses from the last step, in gradient descent method, without its losses will increase not decrease.

The model cannot predict anomalies without learning the correct weights and biases, hence the fine-tuning of weights is done using the iteration over error rate obtained in the previous stage.

The loss function used here is binary cross-entropy with sigmoid [25].

- Optimizer.step:-
  Performs updates in weights and biases based on the gradient and the update rule.
- Loss.backward() :-
  Accumulates the gradient by adding each parameter.

Algorithm 12.2 describes the steps involved in training a deep learning model. The terms used here are from standard Pytorch Documentation [26].

Algorithm 12.2: Training of the multilayer perceptron model using PyTorch

- Initiate loop for the number of epochs
- Declare epoch_loss and epoch_accuracy equal to zero
- Initiate loop for all values of X_in and y_out in train loader
- Load X_in and y_out using .to(device)
- Empty residual gradients
- Update y_pred ← model(X_in)
- Update loss ←a criterion(y_pred, y_out.unsqueeze(1))
- Update accuracy ← model_accuracy(y_pred, y_out.unsqueeze(1))
- Perform backpropagation using loss.backward()
- Perform gradient optimization using optimizer.step()
- Update epoch_loss ← epoch_loss + loss.item()
- Update epoch_accuracy ← epoch_accuracy + accuracy.item()

12.18 Conclusion and Results

In view of the continuing increase in attack types, training a deep learning-based architecture in regular data patterns using the statistical probability density function for normal data is very efficient because the architecture of intrusion detectors varies according to the organization type's own network architecture. Preparing software comprising information about different types of attack is very costly.

However, the multilayer perceptron model can be used for a wide range of network applications, and it does not necessarily require attack information data because it uses deep learning and statistics to find data packets outside normal data patterns (see the model architecture in Figure 12.9).

12.18.1 The Architecture of the Anomaly Detection Model

The key component of a mixed approach using both deep learning and statistics is to use both the probability density function and the multilayer perceptron model and calculate

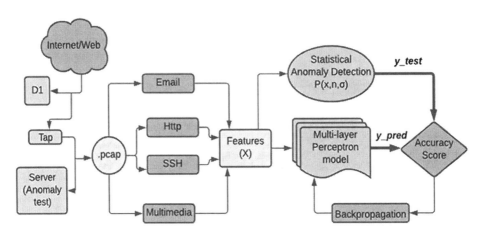

FIGURE 12.9
Architecture of anomaly detection model.

the accuracy score to train the model through backpropagation. The proposed architecture is robust to identify outliers in the normal flow of the network, considering the multilayer perceptrons to learn the underlying pattern distribution. The accuracy results of the architecture are rounded up to the empirical precision of 92.1%.

12.19 Future Work and Research Propagation

The architecture proposed in the chapter uses a multilayer perceptron network to learn the normal behaviour of network packets in the flow of packet transactions between client and server, and to classify them on the basis of the spatial distribution of the network flow.

However, packet transactions in the flow of the network are time-dependent entities; the values of time between each transaction can be extracted using packet-time referencing and the time-relative values of a packet can be stored.

Temporal analysis of a packet specifies a time-dependent feature. It can be used to train a long-short-term memory (LSTM) [27] model which learns the model using backpropagation through time. LSTM models are able to store information over a period of time and can be used to infer a wide set of parameters such as learning rates, input and output biases.

Using a time-series data pattern of normal network flow, the LSTM model can learn the past n number of look-back values of data, and if an anomalous particle is inserted into the network flow it can detect that more efficiently than a multi-layered perceptron network.

However, it has been found that recurrent neural network (RNN) [28] models are insufficient to learn long-term information [29] because they have to travel sequentially through all cells before getting to the present cell. If sequential processing is avoided and units that are better at look-back are found, the attention mechanism [30] for looking for anomalous packets in a time-dependent network flow can largely overcome a vanishing gradient.

Transformer [31] is based entirely on attention mechanism; it is completely different from RNN and CNN and can learn temporal sequential data of feature-value mapping of network packets in a very efficient manner. This system is superior to other architectures such as ANN, CNN, RNN for distinguishing anomalous packets from the distribution function of normal packets in real time.

References

1. Anderson Hiroshi Hamamoto, Luiz Fernando Carvalho, Lucas Dias Hiera Sampaio, Taufik Abrão, and Mario Lemes Proença, Network anomaly detection system using genetic algorithm and fuzzy logic. *Expert Systems with Applications*, vol. 92, pp. 390–402, 2018. SSN 0957-4174.
2. M. H. Bhuyan, D. K. Bhattacharyya, and J. K. Kalita, Network anomaly detection: Methods, systems and tools. *IEEE Communications Surveys & Tutorials*, vol. 16, no. 1, pp. 303–336, First Quarter 2014. doi:10.1109/SURV.2013.052213.00046
3. C. Manikopoulos, and S. Papavassiliou, Network intrusion and fault detection: A statistical anomaly approach. *IEEE Communications Magazine*, vol. 40, no. 10, pp. 76–82, October 2002. doi:10.1109/MCOM.2002.1039860
4. T. Ahmed, B. Oreshkin, and M. Coates, Machine learning approaches to network anomaly detection. In *Proceedings of the 2nd USENIX workshop on Tackling computer systems problems with machine learning techniques*, pp. 1–6. USENIX Association, 2007.
5 K. Rai, M. S. Devi, and A. Guleria, Decision tree-based algorithm for intrusion detection. *International Journal of Advance Networking and Appliances*, vol. 7, no. 4, p. 2828, 2016.
6. N. Farnaaz, and M. Jabbar, Random forest modeling for network intrusion detection system. *Procedia Computer Science*, vol. 89, no. 1, pp. 213–217. doi:10.1016/j.procs.2016.06.047
7. Saurabh Mukherjee, and Neelam Sharma, Intrusion detection using naive Bayes classifier with feature reduction. *Procedia Technology*, vol. 4, pp. 119–128, 2012.
8. Z. Ma, and A. Kaban, K-nearest-neighbours with a novel similarity measure for intrusion detection. Paper presented at *Proceedings of the 13th UK Workshop on Computational Intelligence (UKCI)*. Guildford, UK: IEEE, 2013, pp. 266–271.
9. Leslie F. Sikos, Packet analysis for network forensics: A comprehensive survey. *Forensic Science International: Digital Investigation*, vol. 32, 200892, 2020.
10. Manish Joshi, and Theyazn Hassan Hadi, A review of network traffic analysis and prediction techniques. *arXiv preprint arXiv:1507.05722*, 2015.
11. P. Goyal, and A. Goyal, Comparative study of two most popular packet sniffing tools-Tcpdump and Wireshark. In *2017 9th International Conference on Computational Intelligence and Communication Networks (CICN)*, 2017, pp. 77–81. doi:10.1109/CICN.2017.8319360
12. M. Z. Rafique, and J. Caballero, FIRMA: Malware clustering and network signature generation with mixed network behaviors. In Stolfo S. J., Stavrou A., and Wright C. V. (eds) *Research in Attacks, Intrusions, and Defenses*, RAID 2013. vol. 8145, pp. 144–163. 2013. Springer, Berlin, Heidelberg.
13. Jie Cai, Jiawei Luo, Shulin Wang, and Sheng Yang, 2018. Feature selection in machine-learning: A new perspective. ISSN 0925-2312.
14. David Goldberg, and Yinan Shan, The importance of features for statistical anomaly detection. In *7th USENIX Workshop on Hot Topics in Cloud Computing (HotCloud 15)*, 2015.
15. R. A. Calix, S. B. Singh, T. Chen, D. Zhang, and M. Tu, Cyber security tool kit (CyberSecTK): A Python library for machine learning and cyber security. *Information*, vol. 11, p. 100, 2020.
16. Ioannis Ch Paschalidis, and Yin Chen. Statistical anomaly detection with sensor networks. *ACM Transactions on Sensor Networks*, vol. 7, no. 2, p. 23, 2010, Article 17.
17. C. Manikopoulos, and S. Papavassiliou, Network intrusion and fault detection: A statistical anomaly approach. *IEEE Communications Magazine*, vol. 40, no. 10, pp. 76–82, October 2002. doi:10.1109/MCOM.2002.1039860

18. Fionn Murtagh, Multilayer perceptrons for classification and regression, 1991, ISSN 0925-2312.

19. D. Kwon, H. Kim, J. Kim et al., A survey of deep learning-based network anomaly detection. *Cluster Computing*, vol. 22, pp. 949–961, 2019.

20. Ahmad Javaid, Quamar Niyaz, Weiqing Sun, and Mansoor Alam, A deep learning approach for network intrusion detection system. *EAI Endorsed Transactions on Security and Safety*, vol. 3, no. 9, p. e2, 2016. ISSN 2032-9393.

21. Ambien Fred Agarap, 2018. Deep learning using rectified linear units (ReLU). CoRR. abs/1803.08375. https://dblp.org/rec/journals/corr/abs-1803-08375.bib

22. W. Zhang, Q. Yang, and Y. Geng, A survey of anomaly detection methods in networks. *International Symposium on Computer Network and Multimedia Technology*, 2009, pp. 1–3, 2009. doi:10.1109/CNMT.2009.5374676

23. Ian J. Goodfellow Yoshua Bengio, and Aaron Courville, *Deep Learning*, MIT Press, 2016. https://people.orie.cornell.edu/davidr/or474/nn_sas.pdf

24. W. S. Sarle, Neural networks and statistical models. In *Proceedings of the Nineteenth Annual SAS Users Group International Conference*, Cary, NC: SAS Institute, USA, pp. 1538–1550, 1994.

25. Nguyen Tan, and Scott Sanner, Algorithms for direct 0–1 loss optimization in binary classification. *Proceedings of the 30th International Conference on Machine Learning, PMLR*, 28, no. 3, pp. 1085–1093, 2013.

26. PyTorch Documentation. https://pytorch.org/docs/stable/index.html

27. S. Hochreiter, and J. Schmidhuber, Long short-term memory. *Neural Computation*, vol. 9, no. (8), pp. 1735–1780, 1997. doi:10.1162/neco.1997.9.8.1735

28. Larry R. Medsker, and L. C. Jain, Recurrent neural networks. *Design and Applications*, vol. 5, pp. 64–67, 2001.

29. Razvan Pascanu, Tomas Mikolov, and Yoshua Bengio, On the difficulty of training recurrent neural networks. In *Proceedings of the 30th International Conference on Machine Learning, PMLR* 28(3): 1310–1318, 2013.

30. Attention? Attention!, n.d. https://lilianweng.github.io/lil-log/2018/06/24/attention-attention.html

31. A. Vaswani, N. Shazeer, N. Parmar, J. Uszkoreit, L. Jones, A. N. Gomez, L. Kaiser, and I. Polosukhin, Attention is all you need. *Advances in Neural Information Processing Systems*, pp. 5998–6008, 2017.

13

Intrusion Detection System Using Deep Learning Asymmetric Autoencoder (DLAA)

Arjun Singh

Manipal University, Jaipur, India

Surbhi Chauhan

Jaipur Institute of Engineering Management, Jaipur, India

CONTENTS

13.1 Introduction

While scientific and technological advances are improving network prevention systems, numerous obstacles remain in the form of invading attacks. According to Symantec's 2018 report *Internet Security Threats* [1], the incidence of hostile network incursions increased by 200% in 2018, and the growth rate of assaults against the Internet of Things is even greater at 600%.

DOI: 10.1201/9781003240310-13

The popularity of the modern IoT and continuous use of cloud services has resulted in a significant growth in network data volume, and this trend is expected to continue. A number of new protocols have been added to modern network traffic [2], increasing the difficulties and complexities associated with network security detection systems.

Network protection measures need continuous adjustment to adapt to constantly changing network conditions. The network firewall, which can monitor intrusion activities statically, is one of the most popular measures for defending network security. Intrusion detection systems are employed as a second line of defence for dynamic network security protection, and these can actively protect computer logs and system file changes from signs of attack.

Reviewing log data can detect any changes in the files that could be evidence of an attack, such as unexpected traffic or an undiscovered new attack. There are two types of IDS: information acquired by a single computer system is used by a host-based IDS (HIDS), whereas a network-based IDS (NIDS) collects raw network traffic and analyzes intrusion flags [3]. Machine learning and deep-learning approaches have been utilized in a variety of applications in recent years. Many researchers use them to identify cyber-intrusions, and while measuring methods have been researched, positive alarm accuracy remains low. NIDS is now the biggest cause for concern when it comes to improving generalization performance and efficiency [4].

Most existing intrusion detection systems are still unable to detect new aberrant traffic due to the difficulty in finding valid training data, the duration of training data, and high error rates, among other factors. This strategy necessitates the use of human specialists to filter data, which is both time-consuming and costly [5]. If the detection system relied on human instruction, it would be overloaded. Detection mechanisms should therefore be tweaked so that they can self-learn and detect intrusions with greater accuracy.

In recent years, some shallow K-neighbourhood (KNN, k-nearest neighbour) algorithms [6] and SVM algorithms [7] have shown good performance in intrusion detection. But these learning algorithms also have certain limitations in terms of data samples and ability to express complex functions. To overcome the problem of shallow learning, some researchers have shown that deep-learning algorithms are better than other intrusion methods [8].

This study addresses these issues with a proposal for a new supervised non-symmetric convolutional autoencoder and SVM IDS technique, along with deep-learning and shallow classifiers, for evaluating and discussing data, adapting to changes in modern network traffic, and using the KDD99 data set. Test results suggest that this strategy successfully increases the intrusion detection, efficiency, and detection capabilities of IDS. The following are the principal contributions of this paper.

1. The deep-learning asymmetric autoencoder algorithm (DLAAA) provides non-symmetric data reduction dimension, and solves the deficiencies of convolution neural network and autoencoder [9]. The deep neural network, stacked autoencoder (SAE) and other leading technologies proposed in this article change classification results.

2. There are benefits to combining DLAAA and SVM classification algorithms. Analytic overheads are reduced by employing both deep and superficial learning strategies.

13.2 Literature Survey

Kim [4] suggested a technique for improving intrusion detection using a DNN where 100 units are employed to normalize and preprocess the data. The ADAM optimizer is used to optimize the model. This method was assessed on the KDD-CUP-99 dataset and found to be 100% accurate. At the same time, the author noted that future cyber-defences will include recurrent neural networks and length-periodic memory models. Wang et al. [10] suggested a method based on a hierarchical spatio-temporal characteristic intrusion detection system (HAST-IDS), which first uses a deep convolution neural network (CNN) to learn shallow features, and uses long-term and short-term memory to recall deep features of network learning. The automation process reduces the false alarm rate. This method is applied to the standard data sets DARPA1998 and ISCX2012. Evaluation showed the method is effective in feature learning. Jia [11] built a CNN for IDS exploitation, through which input data was translated into a two-dimensional grayscale image for processing. Ten test data sets were utilized in the testing phase to evaluate the model and the experimental results compared with those of existing intrusion detection systems. CNN-based models have high detection and precision rates. The author shows how intrusion can be detected using a convolutional neural network.

Deep-learning methods have yielded promising results in the field of IDS; some scholars have demonstrated that hierarchical DLA combined with traditional classifiers can outperform traditional classifiers. Shone et al. [5] proposed an asymmetric stacking self-encoding deep-learning algorithm for intrusion detection. The device does feature learning on the data. When the KDD-99 and NSL-KDD data sets were used for a real test, the accuracy rates reached 97.75% and 85.42% respectively. In summary, although some IDSs have achieved good results, there are still many areas requiring improvement. These include the need for supervised learning for data, labour and cost optimization, training time for large amounts of data, and detection accuracy where small samples are unbalanced.

This article uses the unsupervised convolutional autoencoder to process the data line, improving intrusion detection systems for modern network traffic from the perspective of adaptability and detection efficiency, further improving intrusion detection performance.

Deep learning is currently being used in a variety of domains, including medicine [12], autonomous driving [13], image identification [14], and natural language processing [15], with promising results. Many scholars, including Zhao [16], have used deep-learning approaches that combine traditional machine learning with deep-learning approaches for experimental comparison and have found that deep-learning methods outperform traditional ones. Dong et al. [17], comparing classic NIDS techniques with deep-learning methods based on traffic volume and irregular traffic, found that a deep learning technique can increase the accuracy of intrusion detection. They also used a synthetic minority oversampling technique (SMOTE) to prove that the oversampling method can overcome issues related to unbalanced data sets.

13.3 Methodology

13.3.1 Autoencoder

Deep learning algorithms are very popular for the autoencoder, an unsupervised learning method comparable to but capable of outperforming principal component analysis, which

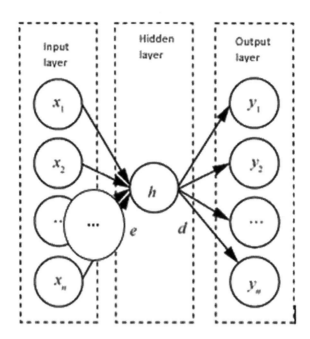

FIGURE 13.1
Autoencoder.

is commonly used in many disciplines. Applications include medical treatment [18] and driving [19], among others. After the data is trained, the autoencoder can try to copy the input to the output as much as possible. There are connections between input layer, hidden layer, and output layer in a deep neural network autoencoder, but no connection between the units of each layer. The autoencoder is basically formed of two parts: (i) the encoder function h = f(x); and (ii) a decoder represented by r = g(h) (h). The autoencoder does not simply enter the data line g(f(x)) = x, as this output is meaningless. Constraints need to be imposed on the autoencoder so that it can only copy the data approximately. It is suitable for feature learning in data since it will automatically select essential characteristics for learning. The output data is compared to the input data, the computed error is reversed, the parameters are modified, and the model is optimized. A typical autoencoder is shown in Figure 13.1.

The role of the hidden layer is to map the high-dimensional information into a low-dimensional form. This stage is the encoding stage. The objective function of the autoencoder is:

$$y = h_{w,b}(x) \approx x \qquad (13.1)$$

h = non-linear assumption

w = weight

b = bias

Throughout the process, it tries to keep the learned input data as similar to the input data x as possible. Inconsistencies are corrected and the reconstruction error function is:

$$L(x, d(e(x)))$$ (13.2)

L is the reconstruction error function, which represents the difference between x and $d(f(x))$.

e is encoding function, and d decoding function.

13.3.2 Convolutional Autoencoder

Convolutional autoencoders combine the advantages of convolutional neural networks with autoencoders to solve the sensitivity of convolutional neural networks to weights and their reliance on large-scale labelling of data. It also addresses the shortcomings of fully linked networks like deep neural networks and autoencoders. For example, the unit completely connected across adjacent layers contains a large number of training parameters. The ideal features in the data can be efficiently retrieved using the convolution kernel, deep learning is utilized to form a deep model structure, and the low-dimensional version of the high-dimensional data is output.

Each subsequent hidden layer reflects more complicated attributes that may minimize the cost of processing, and provide a high degree of accuracy when using several hidden layers to obtain depth. Because each hidden layer's output is utilized as a higher-level input, the first layer is usually used to learn the input data and output the first-order attributes, while the other layer is used to study the second-order features which are associated with the first-order features. The hypothetical model has one convolution kernel. Each convolution kernel is composed of parameters w^l and b^l, and h^l is used to represent the convolution layer.

The convolutional layer is expressed as-

$$h^l = f(x * w^l + b^l)$$ (13.3)

The convolutional operation is represented by the symbol *, while the activation function is represented by f. The following equation can be used to get the h^l feature:

$$y = f(h^l * w^l + c^l)$$ (13.4)

where f, w^l and c^l for the decoder may not be related to the corresponding f, w^l and b^l. The weight of the c^l represents the bias. The data will be entered and lost; data result distance will be compared. The Euclidean distance is used here to reconstruct the error using the gradient drop algorithm and adjusting the parameters to optimize.

$$E = \frac{1}{2}\Sigma(x_i - y_i)2$$ (13.5)

13.4 Proposed Method

Improving the efficiency of intrusion detection is the main objective of modern IDS technology. Therefore, the goal of this research is to establish a fast and efficient intrusion detection system. This chapter introduces an asymmetric convolutional autoencoder-based network intrusion detection system.

13.4.1 Asymmetric Convolutional Self-Encoder

Modern IDS should be designed to improve efficiency. The correct learning structure reduces computational time and overheads which improves model accuracy and efficiency. Asymmetric convolutional self-encoders can be extracted as layered features which can be scaled to accommodate the input of high-dimensional data. Figure 13.2 shows the difference between a balanced convolutional network and an unbalanced convolutional network, where h represents the hidden layer of dimensional reduction, e represents the encoding phase, and d represents the decoding phase.

$x \epsilon R^l$ is the input vector of DLAAA, and the first layer is hidden. The data of the learning input layer is mapped to $h_l \epsilon R^l$, and here l represents the dimension of the vector. Its encoding function can be determined as –

$$h_l = f\left(\omega_i h_{i-1} + b_i\right), i = 1........n \tag{13.6}$$

where, when i is 0, h_0 is x, f represents the activation function, and here the sigmoid activation function is used, n represents the number of hidden layers.

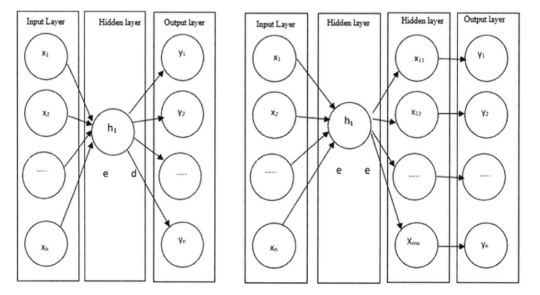

FIGURE 13.2
Comparison of symmetric convolutional autoencoder and asymmetric convolutional autoencoder.

The sigmoid activation function can be expressed as

$$f(t) = \frac{1}{1+e-t} \tag{13.7}$$

The output can be expressed as

$$h_n = f\left(\omega_n h_{n-i} + bn\right) \tag{13.8}$$

The model is back propagated during training to adjust for error, and the reconstruction error of the symmetric autoencoder can be expressed as:

$$E(\alpha) = \sum_{i=1}^{m}\left(x_i - y_i\right)^2 \tag{13.9}$$

where m represents the training sample, and the model adjusts the parameters by minimizing the reconstruction error to achieve a high level of accuracy.

13.4.2 Deep Learning Asymmetric Autoencoder (DLAA)

The asymmetric convolutional autoencoder is not much more accurate than shallow classifiers like KNN [6] and SVM [7] that use a pure asymmetric convolutional autoencoder. Combining deep learning and shallow learning algorithms can increase classification detection accuracy. The SVM algorithm, based on statistical learning methods, is the most widely used machine learning algorithm at present, and is more efficient than traditional methods. Compared with other traditional classification methods, SVM shows good results for small-sample data and high-dimensional data. However, with the arrival of big data, data dimensions are constantly increasing. SVM classification takes a very long time, and there is a high error rate and low true rate. This study therefore developed a new NCAE-NSVM-built algorithm to improve the accuracy of classification detection. Figure 13.3 shows the main architecture of the model, which is a multilayer unsupervised deep neural network divided into three stages.

i. *Preprocessing stage.* The sparse features of the data are merged, digitized and the data is normalized.
ii. *Feature extraction stage.* The asymmetric convolutional autoencoder proposed in Section 13.4.1 is used to extract features from the data.
iii. *Classification stage.* The extracted optimal features are input into a multi-class SVM based on a binary tree for layer-by-layer classification.

The following problems with data and some models are unavoidable:

1. Labelled data resources are very scarce.
2. In many deep learning networks the error function is a highly non-convex function with many local extreme values.
3. Gradient dispersion problems are prone to occur in deep learning networks. These can be resolved using unsupervised asymmetric convolutional autoencoders for feature extraction.

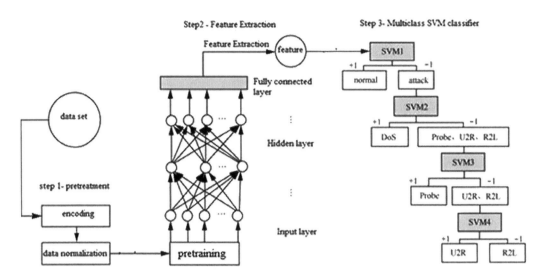

FIGURE 13.3
DLAAA model architecture.

A support vector machine (SVM) is an efficient two-class machine learning algorithm that shows good results for small-sample data and high-dimensional data compared with other traditional classification methods. However, since multi-classification is mostly required, the training time for SVM is usually very long. In our study, the extracted features were input into a multi-class SVM classifier based on a binary tree structure. *k-1* SVM classification is needed to detect *k* classification data. We therefore conducted four SVM detection steps for five types, among which the Gaussian kernel function is used to solve the classification problem of non-linear samples. In this process, SVM only outputs +1 and −1. The multi-class classification steps are as follows.

a. Input the obtained features into SVM1. SVM1 checks whether the data is a normal or attack type. If it is an attack type, the SVM output is −1 and the attack-type data is input into SVM2, otherwise the output of SVM is +1.
b. After SVM2 receives the output data of SVM1, it checks whether the data obtained is from DOS or U2R and R2L probes. If it is U2R or R2L, the SVM output is −1 and this type of data is input into SVM3.
c. After SVM3 receives the output data of SVM2, it checks whether the detected data probe is U2R or R2L. If it is a U2R or R2L attack type, SVM output is −1 and this type of data is input into SVM4 for classification, otherwise, the SVM output is +1.
d. After SVM4 receives the output data of SVM3, it checks whether the data obtained is U2R or RL. If it is U2R, the SVM output is +1, otherwise, the SVM output is −1.

The number of hidden layers in the model, the output feature mapping dimension of each layer and the parameters (kernel function and penalty factor) in the SVM are all optimal parameters obtained by ten-fold cross-validation.

13.5 Model Complexity and Timeliness

The asymmetric convolutional autoencoder is primarily responsible for the computational complexity of the DLAAA intrusion detection suggested in this paper. $O(M^2/K^2IT)$ is the time complexity of a single convolutional layer, where M is the side length of each convolutional kernel output feature map, K is the side length of each convolutional kernel, and I is the length of each convolutional layer. T represents the number of output channels of each convolutional layer. The size of M is determined by the size of the input data L, the size of the convolutional kernel K, *padding*, and the step size of the convolution is determined by *Stride:* $M = \dfrac{L - k + 2 \times paddng}{Siride}$.

The overall complexity of the model can be calculated from the single-layer complexity degree is $0 \left(\displaystyle\sum_{i=1}^{N} M_i^2 k_i^2 I_i T_i \right)$ where N means that model has a number of convolutional layers, I represents the number of convolutional layers of the model. The model's space complexity is mainly determined by the size of the convolutional kernel K. The number of output channels $I_i T_i$ and the depth of the neural network determine N. It has nothing to do with size of the data and can be expressed as $0 \left(\displaystyle\sum_{i=1}^{N} k_i^2 I_i T_i \right)$.

Analyzing the time and space complexity of the model shows that this intrusion detection system has high realizability and can run on any operation system. As regards modern network detection systems against unknown intrusion events, this model is strongly competitive and efficient, as it does not require a large amount of manually labelled data or preprocessing.

13.6 Experiment

The experimental process described in this paper is shown in Figure 13.4. First, KDD99 data is normalized. The standard data set obtained is input into the convolutional autoencoder for feature extraction, and the multi-class SVM classifier is used for training and testing. The deficiencies of the model are analyzed in respect of accuracy, false negative and false positive rates and the optimized model is obtained. The proposed model is evaluated by comparing its results with the performance of other machine-leaning algorithms.

13.7 Experimental Data

The KDD 99 dataset used in this study consists of network connection and system audit data collected by the United States Air Force (USAF) over a period of nine weeks. It simulates various user types, network traffic and attack methods in a real network environment.

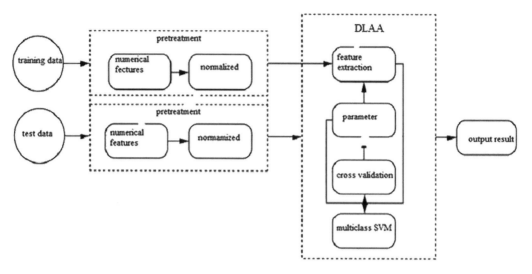

FIGURE 13.4
Experimental process.

TABLE 13.1

Number of Records in KDD99 Data Set

Attack Type	Training Set	Test Set
Normal	97280	60595
DoS	391460	23085
Probe	4110	4167
U2R	55	229
R2L	1126	16190

Its training data includes 50,000 single items of connection data; the test data contains one million network connection data. The training data contains a total of 494,021 records, only 11% of which are used for training. The dataset includes five types: normal, DoS, R2L, U2R, probe. There are 39 types of attack, 22 of which appear in the training set, with the other 17 unknown types appearing in the test set. Detailed data information is provided in Table 13.1.

13.8 Data Preprocessing

Each link in the KDD 99 data collection has 41 features, many of which are irrelevant or non-numerical. The data collection has to be preprocessed by converting it into features that the model can accept. Numerical features and normalizing are the two key aspects of preprocessing.

TABLE 13.2

Experimental Variable Parameters

Variable Name	Variable Size
1st layer	$20 \times 20 \times 128$
2nd layer	$10 \times 10 \times 128$
3rd layer	$5 \times 5 \times 32$
4th layer	$3 \times 3 \times 32$
5th layer	$5 \times 5 \times 64$
6th layer	$5 \times 5 \times 64$
7th layer	$3 \times 3 \times 64$
8th layer	$3 \times 3 \times 128$
9th layer	$2 \times 2 \times 128$
Fully connected layer	512
Convolutional kernel size	3×3
Learning rate	0.001
Convolutional kernel step size	1
Parameter of kernel function	.0001
Penalty function	1.000

13.8.1 Numerical Features

Neural network training requires the use of numerical features. In the preprocessing stage, it is necessary to convert non-numerical features into numerical features. There are three protocol types in the KDD99 dataset: 70 service symbol values and 11 label symbol values are all non-numerical. One-hot encoding is used to establish corresponding numerical mapping (such as TCP = [0-0-0], UDP[0-1-0] and ICMP[1-0-0] conversion.

13.8.2 Normalization

Because there are some discrete or connected values in the KD99 data, and their ranges are different, the data is not comparable between dimensions. The normalization method maps the numeric attributes between [0, 1] as follows:

$$Z_K = \frac{Z - Z_{min}}{Z_{max} - Z_{min}} \tag{13.10}$$

Z is the value of a certain dimension in the data, Z_{min} is the minimum value of the dimension, Z_{max} is the maximum value of the dimension, and Z_k is the final normalization data (Table 13.2).

13.9 Experimental Environment and Parameters

For this experiment following hardware and software is used:

Processor = Intel ® core i3-2310M CPU @2.1GHz

RAM = 3GB

HDD =1 TB

OS = Windows 8.1 Pro 32-bit

Programming language = Python 3.5

13.10 Evolutional Index

The assessment indicators used to assess the model's performance are: accuracy (4), precision rate (P), result rate (R), false alarm rate (F), and false alarm rate (F) (M). The overall performance is measured by the accuracy rate, false alarm rate, and false alarm rate, while the comparison across models is measured by the accuracy rate and recall rate. Table 13.3 shows the confusion matrix.

Accuracy refers to the number of samples that are correctly classified by the classifier. This ratio of this total is calculated as:

$$A = \frac{TP + TN}{TP + TN + FN} \tag{13.11}$$

The false negative rate is the sample and the actual category predicted by the classifier. The formula for the ratio of all samples with 0 is as follows:

$$M_R = \frac{FP}{FP + TN} \tag{13.12}$$

False alarm is defined as

$$M_R = \frac{FN}{FN + TN} \tag{13.13}$$

The recall rate means that the type with the predicted category 1 is classified as 1. This is calculated as

$$R = \frac{TP}{TP + FP} \tag{13.14}$$

TABLE 13.3

Confusion Matrix

		Predict	
Confusion Matrix		**1**	**0**
actual	1	TP	FN
	0	FP	TN

TP: Class 1 is correctly predicted
TN: Category 0 is correctly predicted to be category 0
FP: Category 0 was incorrectly predicated as category 1
FN: class 1 is incorrectly predicated to be class 0

Accuracy refers to the probability that the prediction is correct in the sample predicted to be 1, and its calculation formula is as follows:

$$P = \frac{TP}{TP + FN}$$ (13.15)

13.11 Simulation Experiments and Result Analysis

Experiment 1:   The effect of the number of model layers on intrusion detection

- The detection of the number of layers in the deep neural network has important implications. This paper analyzes the result of asymmetric convolution auto-encoding. Their detection indicators are accurate rate, false positive rate, and false positive rate. As shown in Table 13.4, there are six different asymmetric convolutional autoencoder hidden layers: 5, 6, 7, 8, 9, and 10. When the top-down network layer increases, the data detection results are hidden to some extent; as the number of hidden layers increases, the false positive rate also increases. As the underreporting rate decreases, it becomes more conducive to convert high-dimensional data to low-dimensional data, and both model detection accuracy and classification speed can be improved. In the 9-layer structure of the neural network the test results are optimal, with an accuracy rate of 97.71%. In the other model structures, false positives were 3.11%. The underreporting rate is 7.22%, which is superior to other models. The 9-tier convolution autoencoder model structure for intrusion detection systems was therefore selected.

Experiment 2:   The effect of number of iterations on performance

- The number of iterations determines whether the model learns the features in the data. This is a black-box process, so it is necessary to adjust the appropriate parameter experimentally. Figure 13.5 shows the effect of the number of iterations on detection loss. At fewer than ten iterations, the false negative rate and false positive rate are high, mainly because the neural network has not learned all the data characteristics. When the number of iterations is between ten and 20, the model learns all data features, the accuracy rate is maintained at a high level, and the false alarm rate remains low. When the number of iterations reaches more than 20, the accuracy rate decreases and the false alarm rate rises. So the number of iterations is kept between ten and 20.

TABLE 13.4

The Influence of Asymmetric Autoencoder on Detection Results

Number of Layers	Accuracy	False Alarm Rate	False Negative Rate
5	88.30%	4.75%	9.85%
6	89.46%	3.97%	9.73%
7	96.67%	3.85%	8.35%
8	96.15%	3.48%	8.47%
9	97.93%	3.16%	7.36%
10	95.32%	3.27%	8.13%

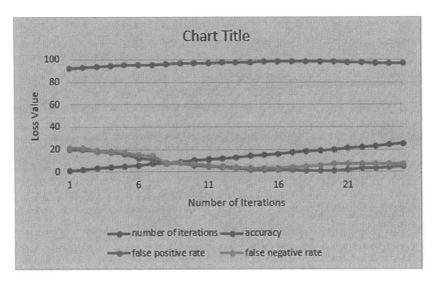

FIGURE 13.5
Number of iterations vs loss values.

TABLE 13.5

Comparison of Different Algorithms

Attack type	CHI-SVM			SNDAE			DBN			DLAAA		
	A	P	R	A	P	R	A	P	R	A	P	R
Normal	96.10%	99.50%	99.60%	99.49%	100.00%	99.49%	99.51%	94.49%	99.65%	99.52%	100%	99.85%
DoS	98.87%	99.80%	99.90%	99.79%	100.00%	99.79%	99.65%	98.74%	99.65%	99.65%	99.63%	99.66%
Probe	95.80%	99.20%	99.20%	98.74%	100.00%	98.74%	14.19%	86.66%	14.19%	99.85%	99.74%	99.73%
R2L	76.92%	99.20%	98.70%	9.31%	100.00%	9.31%	89.25%	100.00%	89.25%	22.35%	99.30%	86.25%
U2R	96.37%	89.50%	73.90%	0.00%	0.00%	0.00%	7.14%	38.46%	7.14%	9.56%	21.58%	43.53%

A = Accuracy, P = Precision, R = Recall

Experiment 3: Performance comparison with other methods

- Our results were compared with three other methods. References [5, 20, 21] use the same evaluation criteria as in this paper: accuracy, precision, recall, and time. Table 13.5 shows the classification of each algorithm for each type of sample in the dataset. Figure 13.6 compares the test time of each algorithm. For CHI-SVM [20], the classification of large samples in the dataset is not as good as in the present paper, but for small samples U2R and R2L are better. A comparison with the SNDAE algorithm shows that classification for small samples is not good, and recall and accuracy of U2R are also zero. During high traffic on the network, some attacks are not detected, which can cause a huge threat to network security.

- The proposed DLAA algorithm combines the advantages of a convolutional neural network and autoencoders. This solves the sensitivity of convolutional neural networks to weights and their dependence on large data. It also avoids some shortcomings of fully connected networks such as DBN and autoencoder.

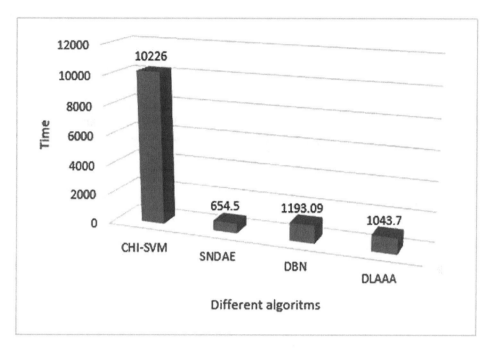

FIGURE 13.6
Test time of different algorithms.

Experiments showed that the proposed algorithm achieved good results in accuracy, improved training time and performed better in the case of false negative and false positive alarms.

13.12 Conclusion and Future Work

This chapter reported an extensive investigation of deep learning algorithms in the field of IDS. With today's exponential increase in network traffic, many new protocols have emerged. Most existing IDSs cannot detect the threats due to their legacy methods. Existing systems also have limitations in terms of scalability, detection efficiency, and recall. With these parameters in mind, a good algorithm (DLAAA) is required to mark and preprocess the data. This model also solves the problems of autoencoders and neural networks. Neural networks are very sensitive to weight and rely on large-scale training data. Our experiments showed that the overall accuracy of the model is 97.96%, and the training time also very low compared to other algorithms. We will continue to conduct in-depth research in this area with the aim of adding more classification techniques for intrusion detection.

References

1. Symantec. (2018). Internet security threat report. istr-23-2018-en (broadcom.com)
2. Liao, H. J., Lin, C. H. R., Lin, Y. C., & Tung, K. Y. (2013). Intrusion detection system: A comprehensive review. *Journal of Network and Computer Applications, 36*(1), 16–24.

3. Chowdhury, M. M. U., Hammond, F., Konowicz, G., Xin, C., Wu, H., & Li, J. (2017, October, 19–21). A few-shot deep learning approach for improved intrusion detection. In *2017 IEEE 8th Annual Ubiquitous Computing, Electronics and Mobile Communication Conference (UEMCON)*, Columbia University, New York City (pp. 456–462). IEEE.

4. Kim, J., Shin, N., Jo, S. Y., & Kim, S. H. (2017, February). Method of intrusion detection using deep neural network. In *2017 IEEE International Conference on Big Data and Smart Computing (BigComp)* (pp. 313–316). IEEE.

5. Shone, N., Ngoc, T. N., Phai, V. D., & Shi, Q. (2018). A deep learning approach to network intrusion detection. *IEEE Transactions on Emerging Topics in Computational Intelligence*, 2(1), 41–50.

6. Dave, D., & Vashishtha, S. (2013). Efficient intrusion detection with KNN classification and DS theory. In *Proceedings of All India Seminar on Biomedical Engineering 2012 (AISOBE 2012)* (pp. 173–188). Springer, India.

7. Aburomman, A. A., & Reaz, M. B. I. (2016). A novel SVM-kNN-PSO ensemble method for intrusion detection system. *Applied Soft Computing*, 38, 360–372.

8. Hou, S., Saas, A., Chen, L., & Ye, Y. (2016, October). Deep4maldroid: A deep learning framework for android malware detection based on linux kernel system call graphs. In *2016 IEEE/WIC/ACM International Conference on Web Intelligence Workshops (WIW)* Omaha, NE, USA (pp. 104–111). IEEE.

9. Ji, S., Ye, K., & Xu, C. Z. (2020, September). A network intrusion detection approach based on asymmetric convolutional autoencoder. In *International Conference on Cloud Computing* (pp. 126–140). Springer, Cham.

10. Wang, W., Sheng, Y., Wang, J., Zeng, X., Ye, X., Huang, Y., & Zhu, M. (2017). HAST-IDS: Learning hierarchical spatial-temporal features using deep neural networks to improve intrusion detection. *IEEE Access*, 6, 1792–1806.

11. Liu, Y., Liu, S., & Zhao, X. (2017). Intrusion detection algorithm based on convolutional neural network. In *DEStech Transactions on Engineering and Technology Research*, (iceta).

12. Shen, D., Wu, G., & Suk, H. I. (2017). Deep learning in medical image analysis. *Annual Review of Biomedical Engineering*, 19, 221–248.

13. Liu, H., Taniguchi, T., Tanaka, Y., Takenaka, K., & Bando, T. (2015, June). Essential feature extraction of driving behavior using a deep learning method. In *2015 IEEE Intelligent Vehicles Symposium (IV)* (pp. 1054–1060). IEEE.

14. Grm, K., Štruc, V., Artiges, A., Caron, M., & Ekenel, H. K. (2018). Strengths and weaknesses of deep learning models for face recognition against image degradations. *Iet Biometrics*, 7(1), 81–89.

15. Gardner, M., Grus, J., Neumann, M., Tafjord, O., Dasigi, P., Liu, N., ... & Zettlemoyer, L. (2018). Allennlp: A deep semantic natural language processing platform. *arXiv preprint arXiv:1803.07640*.

16. Zhao, R., Yan, R., Chen, Z., Mao, K., Wang, P., & Gao, R. X. (2019). Deep learning and its applications to machine health monitoring. *Mechanical Systems and Signal Processing*, 115, 213–237.

17. Dong, B., & Wang, X. (2016, June). Comparison deep learning method to traditional methods using for network intrusion detection. In *2016 8th IEEE International Conference on Communication Software and Networks (ICCSN)* Beijing, China (pp. 581–585). IEEE.

18. Xu, J., Xiang, L., Liu, Q., Gilmore, H., Wu, J., Tang, J., & Madabhushi, A. (2015). Stacked sparse autoencoder (SSAE) for nuclei detection on breast cancer histopathology images. *IEEE Transactions on Medical Imaging*, 35(1), 119–130.

19. Dong, W., Yuan, T., Yang, K., Li, C., & Zhang, S. (2017). Autoencoder regularized network for driving style representation learning. In *Proceedings of the 26th International Joint Conference on Artificial Intelligence* (acm.org) *arXiv preprint arXiv:1701.01272*.

20. Thaseen, I. S., & Kumar, C. A. (2017). Intrusion detection model using fusion of chi-square feature selection and multi class SVM. *Journal of King Saud University-Computer and Information Sciences*, 29(4), 462–472.

21. Alrawashdeh, K., & Purdy, C. (2016, December). Toward an online anomaly intrusion detection system based on deep learning. In *2016 15th IEEE International Conference on Machine Learning and Applications (ICMLA)* (pp. 195–200). IEEE.

22. Tavallaee, M., Bagheri, E., Lu, W., & Ghorbani, A. A. (2009, July). A detailed analysis of the KDD CUP 99 data set. In *2009 IEEE Symposium on Computational Intelligence for Security and Defense Applications* Anaheim, CA, USA (pp. 1–6). IEEE.

Index

9781032146416